*How to Enjoy
Mathematics
with
Your Child*

How to Enjoy Mathematics with Your Child

Nancy Rosenberg

STEIN AND DAY/*Publishers*/New York

First published in 1970
Copyright © 1970 by Nancy Rosenberg
Library of Congress Catalog Card No. 70-104628
All rights reserved
Published simultaneously in Canada by Saunders of Toronto Ltd.
Designed by Bernard Schleifer
Printed in the United States of America
Stein and Day/*Publishers*/7 East 48 Street, New York, N.Y. 10017
SBN 8128-1293-X

Contents

To my students at the Riverdale Country School,
who have taught me so much.

Introduction

"BUT WHY do we have to learn this?" Mathematics teachers probably hear this question more often than anyone else does. Today's students are required to learn more mathematics, faster, than any students in history. They are under tremendous pressure to succeed, although most of them are no more comfortable with the subject than their parents were. No wonder their question is so often asked out of resentment, or even desperation.

I think these students deserve an answer, and in my own classroom, over a period of ten years, I have been trying to find one for them. I started by repeating exactly what my own teachers told me. "The study of mathematics," I said, "develops orderly thinking habits and a disciplined mind." This sounded good, and it silenced the students, but I don't think it ever convinced them.

Then I began talking about the satellites that relay pictures from one continent to another and computers that do everything from designing aircraft to predicting the weather. "Surely," I told my students, "educated people should have some understanding of the mathematics that makes all this possible. And facility with numbers is a practical necessity. Without it we cannot balance our budgets, compute our taxes, or understand what we read in the newspapers."

This statement, like the first one, was quite true. But neither of

I

them had much appeal for the students, because neither related
directly to their experience. Both carried an implicit, "Take my
word for it," and this is something today's young people are in-
creasingly reluctant to do.

By this time I had been teaching long enough to realize that one
group of students never asked why they had to study mathematics.
They enjoyed it so much that the question never even occurred to
them. These were the children who found problem solving an exciting challenge. "Don't tell us," they would say. "Let us figure it out
for ourselves." While the class was being taught one method of
solution, they were working out another in the margin. No one had
to tell them that mathematics helped develop their minds; they
discovered that for themselves. And its applications to the world
around them were a source of endless fascination.

With these children in mind, I formulated another answer to
the perennial question. It didn't sound as impressive as the others,
but it seemed more satisfying to the students. It went something
like this: "Our educational system is set up in such a way that you
have to do reasonably well in mathematics in order to do much of
anything else. The colleges demand it. Nearly all college-bound
students take mathematics through their junior year of high school;
their college choices are severely limited if they don't. And the
courses have been upgraded repeatedly. This situation reflects an
appreciation on the part of educators of the tremendous value and
importance of mathematics, and we hope that your study of it will
instill some of this same appreciation in you. But like it or not,
mathematics is going to be part of your life for a number of years,
and you might as well enjoy it. The most successful students are the
ones who do. And mathematics, perhaps more than any other
subject, can be a great deal of fun."

This explanation appealed to the students because it was honest.
Most of them *do* take mathematics because it is required. On the
other hand, those of us who teach it hope that our audience will be
appreciative even though it is captive. And I have never seen a
child really appreciate mathematics unless he enjoyed it, too.

Why do some children take to mathematics so naturally and

easily while others find it such a struggle? Ability has a lot to do with it, of course. But a number of other factors are also important. A recent study compared the mathematics achievement of students in twelve countries, including the United States, and American educators were far from pleased with the results. The Japanese students far outdid their American counterparts. So did the students in England, Belgium, France, and Sweden.

Experts consider American attitudes toward mathematics largely to blame for this poor showing. The subject is accorded much less respect in this country than it is in many of the others. Where excellence is not valued and expected, it is apparently not attained. Many parents who set high standards for their daughters in every other subject have said to me, "All we want her to do is pass." Very often they are lucky if she does.

It is not surprising that girls do not do as well in mathematics as boys, for whom the expectations are higher. What is interesting is the fact that the girls' performance is found to improve in co-educational classes where they find the stimulation and incentive to make the most of their abilities.

When it comes to reading, many parents provide this kind of stimulation at home. Most children are introduced to the world of books long before they sound out their first word. When the time comes for them to explore it on their own, they can scarcely wait to begin. But parents who read to their children by the hour rarely think of playing number games with them. Any attitudes they do instill toward mathematics are likely to be negative ones. "Poor Sally," many a mother has said to me, "no wonder she has such trouble with math. Neither her father nor I could ever do it either."

I have known a number of Sallys and some of their brothers. Often, they are studious and intelligent. Sometimes they are the joy of the English department. But none of them has ever regarded mathematics in the same light as the other subjects. At best they find it a dreary nuisance; at worst it is the bane of their existence. In its extreme form, this attitude has come to be known as a "math block." If it persists into the high school years, there is very little that can be done about it. The child is at a permanent disadvantage, not only in school but throughout his life.

In most cases, this situation is as unnecessary as it is disturbing. Any child who can appreciate a good book can enjoy the excitement of making mathematical discoveries. Properly directed, he should be just as successful in mathematics as he is in his other subjects. And this direction need not be confined to the classroom. Parents, even Sally's, can do a great deal to develop positive attitudes toward mathematics.

This book is a collection of mathematical activities and recreations that parents can enjoy with their children. Although most require no more than a knowledge of simple arithmetic, it is difficult to designate the most suitable age group for each of them. The best judge of the suitability of any particular activity for any particular child is probably the child himself.

Much of the subject matter is related to material that is included in the school curriculum, but it is not designed to teach any particular topic or skill. Rather, it is intended to show that mathematics can be a source of pleasure and fascination both inside the classroom and out.

I

Arithmetic Without a Pencil

REMEMBER the arithmetic book you used as a child? Most of it was just page after page of addition, subtraction, multiplication, and division problems. If there were any pictures at all, they were probably black and white photographs of girls doing their mothers' marketing and boys helping their fathers measure boards. No one bothered to explain such things as borrowing and carrying; very likely the teacher didn't understand them herself. They worked, and that was all that mattered.

Today's books, by contrast, are varied and colorful. They are illustrated with Venn diagrams, number lines, and pictures of enormous computers. Much of the material they present is frighteningly unfamiliar, and very little of it looks as though it would be much help with anything as practical as marketing or measuring.

The last hundred years have seen unprecedented progress in mathematics, but until recently very little of it was reflected in the classroom. School children were being taught much the same material, in much the same way, as their parents and even their grandparents. It was late in the 1950's when the first group of teachers and mathematicians got together to revise the mathematics curriculum and bring it more in line with current thinking in the field. The results, in the form of new math, are well known to parents everywhere.

New math has brought changes in both content and approach.

5

Set theory, the nature of our number system, and the properties of numbers are being taught in the early grades. Children are encouraged to make mathematical discoveries on their own and to understand the reason for everything they do. When the time comes to borrow and carry, they are shown exactly what it means and why they must do it.

The new approach is paying off in a number of ways. Instead of learning by rote and drilling for hours, children are beginning to discover the power and beauty of mathematics. They are enjoying it more than they ever have before. But the new math hasn't made the old math any less important. The commutative and associative laws can help a child to understand the *why* of multiplication, but he still won't be able to get the right answers unless he knows the tables.

This is one place where parents can help. No matter what the state of your mathematical knowledge, you can work with your child on the basic facts of arithmetic. And with a little extra effort you should be able to give him a valuable skill that is almost never taught in school, the ability to do simple problems in his head.

If you aren't accustomed to doing arithmetic mentally yourself, you may have to practice a little before you are ready to work with your child. But luckily it's the simple problems that present themselves most often, and no special talent is needed to solve them mentally. Any child who can do elementary arithmetic on paper can do it in his head, provided he is taught early enough.

Unfortunately, most school children are forced to write down every problem they do, step by step. This may be to the teacher's advantage, but it isn't always to the student's. Arithmetic can be tedious and time consuming. Shortcuts are all to the good, and the biggest shortcut of all is doing the problems mentally. But once a student is accustomed to writing everything out, it can be extremely difficult for him to put down his pencil.

The sooner you start your child on mental arithmetic, the better. Like all good habits, this one is best learned early. Once he knows the numbers from one to one hundred, he can add fifty and thirty as easily as he can add five and three. He will begin to appreciate the power of numbers when he sees how easily his knowledge can be extended, and two-digit problems will reinforce his knowledge of the basic facts of addition while preparing him for more difficult combinations.

Practice until his answers become automatic. Then, when 50 + 30 is second nature to him, try 50 + 32. This will introduce him to a two-step problem. He must learn to break up the 32 into 30 + 2, add the 50 to the 30 as before, and then tack on the 2.

It's just a short step from 50 + 32 to 53 + 32. The easiest way to handle this kind of problem is to add the tens and the units separately. Think of 53 as 50 + 3 and 32 as 30 + 2. Now numbers that are to be added can be shuffled and grouped in any way at all. There are special names for the shuffling and grouping properties; they are called the commutative and associative laws of addition. For our purposes it is most convenient to think of (50 + 3) + (30 + 2) as (50 + 30) + (3 + 2). This leads to 80 + 5, and finally to 85.

Problems like these require a lot of practice, but you probably have more time for this than you realize. Give your child examples to do when you're riding in the car together or waiting on line at the supermarket. Do it in the same spirit as you would play a game of tick-tack-toe, always taking your cue from him. If he seems enthusiastic, give him more problems. If he'd rather play ball, let him. He'll learn the addition and multiplication tables sooner or later, with or without you. You'll be accomplishing something far more important if you can teach him to enjoy arithmetic and to do it mentally while he's still flexible enough to learn how.

The most difficult part of addition is carrying. A child who is thoroughly at ease with 53 + 32 will probably balk at 57 + 38. 50 and 30 make 80 all right, but 7 and 8 are 15. What is to be done with that extra digit? It must be carried over to make the 80 a 90, of course, and this can be confusing, expecially when the work is being done mentally. The thought process becomes (50 + 7) + (30 + 8) equals (50 + 30) + (7 + 8), which is 80 + 15, which is 80 + (10 + 5), which is 90 + 5, which is 95. Far too much trouble.

Now the easiest way to deal with a complicated problem is to turn it into a simple one. Can we do that here? Is there some way we can avoid the carrying altogether?

Suppose we were to add 57 to 40 instead of to 38. The answer would be much easier to find, but it would turn out to be 97 instead of 95, 2 too much. This extra 2 came in when we changed the 38 to 40. Is there any way we can compensate for it?

There are several. We can simply subtract 2 from the 97. Or we can take 2 away from the 57 before we add it. Having added and subtracted 2, we will have left the sum unchanged. $57 + 38$ adds up to the same thing as $55 + 40$, but where the first problem involves carrying, the second avoids it completely. We could have accomplished the same thing by changing the 57 to 60 and the 38 to 35. In this case we would have added and subtracted 3, again leaving the sum unchanged. $60 + 35$ gives the right answer too.

These same techniques can be used to add columns of figures. Here's a typical third or fourth grade problem:

$$
\begin{array}{r}
12 \\
23 \\
8 \\
27 \\
+\ 6 \\
\hline
\end{array}
$$

Most children would tackle it by adding the numbers in the units column, one by one, and getting 26. Then they would "put down the 6 and carry the 2," adding it to the tens digits to get 7. This gives a sum of 76.

For the child who can add two digit numbers in his head, there's an easier way. He can add both columns at the same time, like this: $12 + 23$ is 35, $35 + 8$ is 43, $43 + 27$ is 70 and $70 + 6$ is 76. To speed things up a bit, he could just look at the numbers and think 35, 43, 70, 76.

This process should take only seconds, but with a little ingenuity the problem can be solved even faster. There's no need to add the numbers in the order they happen to be given; any other will work just as well. Suppose we were to group the 12 with the 8 and the 23 with the 27. This gives us $20 + 50$, to which we need only add the 6 to get 76.

Numbers like 20 and 50 are easy to work with because they end in zero. 12 and 8 add up to a number that ends in zero because $2 + 8 = 10$. It's the same story with 23 and 27; 3 and 7 are 10 too. So are 1 and 9, 4 and 6, and 5 and 5. Whenever the numbers in a sum can be paired up this way, it makes the work that much easier.

One thing you find out if you teach math long enough is that different people think in different ways. A method that is comfortable and natural for one student can be awkward and difficult for another. So the best approach is to explore the alternatives and then let your child make his own choice. This has another advantage too. There's no better way to understand a problem thoroughly than to study it in the light of all its different methods of solution.

See what you can do with this one:

$$
\begin{array}{r}
14 \\
13 \\
5 \\
12 \\
+\ 7 \\
\hline
\end{array}
$$

13 and 7 make 20, but none of the three remaining numbers can be combined to give a number that ends in zero. $14 + 5$ is 19 though, just 1 less than 20. Suppose we were to think of the 12 as $1 + 11$. Then we could combine this 1 with the $14 + 5$ to get another 20. This would give $20 + 20 + 11$, or 51.

There are other ways to do the problem too. 5 and 7 are 12; so the last three numbers taken together equal two 12's. 13 and 14 equal two 12's too, with 3 left over: 1 from the 13 and 2 from the 14. So altogether there are four 12's + 3, or 51. This may sound complicated, but with enough practice the four 12's, or some similar grouping, can be seen at a glance.

How far your child is able to go with mental arithmetic will depend on his imagination and ingenuity and how well he is taught to use them. Writing problems out may be tedious, but if you know the method and carry it out correctly, you can get the right answer with very little thought. In mental arithmetic, different problems require different methods, and thinking is the key to everything.

Three-digit numbers can be added the same way as two-digit numbers, but they're more difficult to remember. Take 578 and 294. 78 + 94 isn't hard; it can be changed to 80 + 92, which is 172. But it's awfully easy to forget the 500 and the 200 while you're working this out. A better method might be to add 6 to the 294 to turn it into 300. The 578 would then be lowered by 6 to 572, and 300 + 572 gives the answer in one step.

This trick wouldn't have worked if 294 hadn't been so close to 300. Suppose the problem had been 578 + 269. We might have handled it by combining 78 and 69 first and then adding that sum to 700. Or we might have used another method. 578 is just 2 less than 580, and 269 is 1 less than 270. If we can add 58 to 27, we can add 580 to 270, and this sum is just 3 more than the one we want.

All of this may seem like a lot more trouble than it's worth, but your child will probably find it easier than you do. In fact, he may even enjoy the challenge posed by each new problem. If he seems to be developing the ability to remember longer numbers and manipulate them mentally, it's probably worth some time and effort on your part. If he balks, forget about it. The smaller numbers come up a lot more often anyway.

Addition and subtraction are really two ways of looking at the same thing. Since 3 and 4 make 7, 7 − 3 is 4 and 7 − 4 is 3. If a child can add, he can subtract. And if he can do addition mentally, he ought to be able to do subtraction that way too.

7 − 3 is 4, so 70 − 30 is 40 and 700 − 300 is 400. Practice with easy combinations like these until subtraction becomes as automatic as addition. Then try a problem like 74 − 30. The fact that 30 is a multiple of 10 makes this one very simple too. All multiples of 10 end in zero, and zero is the easiest number to subtract because it always leaves the number above it unchanged. 70 − 30 is 40 as before, and zero from 4 is 4. Answer: 44.

Even if the 30 had been a 33, the problem could have been done mentally without too much difficulty. But sooner or later there will be a problem like 74 − 39, and this is where the trouble starts. To avoid it, it is necessary to "borrow."

Many of today's parents may remember being taught to subtract this way:

$$7^14$$
$$-\,^4\cancel{3}\,9$$
$$\overline{3\ 5}$$

They were really adding ten to both numbers. 74 became $70 + 14$ and 39 became $40 + 9$. This method works because the difference between two numbers remains unchanged when the same thing is added to both of them.

Your child's subtraction problems are more likely to look like this:

$$^6\cancel{7}^14$$
$$-\ 3\ 9$$
$$\overline{3\ 5}$$

He isn't adding anything; he's just writing 74 as $60 + 14$.

Both of these methods work well on paper, but when subtraction is done mentally, it's easier to avoid borrowing altogether. How can this be done? The simplest way is by changing the 9 in the 39 to a zero, and this can best be done by adding 1 to it and to the 74 as well. Then, instead of taking 39 from 74, we take 40 from 75 and get the same answer painlessly.

This trick makes it easy to subtract any two-digit number from another without borrowing. If the smaller number has a higher units digit than the larger one, raise or lower it to the nearest multiple of ten. Then change the larger number by the same amount. The units digit of the answer is always the units digit of the larger number after it has been changed. The tens digit of the answer is the difference between the tens digits of the two new numbers. Suppose the problem had been $41 - 27$. The multiple of ten nearest 27 is 30, so both numbers must be raised by 3. Instead of subtracting 27 from 41 we subtract 30 from 44. The units digit of the answer is 4. The tens digit is $4 - 3$, or 1. Answer: 14.

This method works just as well when two digits are being subtracted from three. If the problem were $142 - 88$, we'd add 2's to get $144 - 90$. The units digit of the answer is 4, the tens digit is $14 - 9$, and the answer is 54.

The same method can be extended to pairs of three-digit numbers, but, as in addition, some combinations are harder than others. A problem like $275 - 194$ is just as easy as its two-digit counterpart. Adding 6 to both numbers gives $281 - 200$, and this is clearly 81.

$321 - 177$ is harder. The best way to do it is to change it twice, first to $324 - 180$ by adding 3's, and then to $344 - 200$ by adding 20's. Then the answer, 144, can be found in one step.

Some problems lend themselves to special methods. Take $275 - 177$. Here we could add 3's to get $278 - 180$ and 20's to get $298 - 200$, but it would be much easier to think of it another way. If the problem were $275 - 175$, the answer would be 100. But since we are taking away 2 more than 175, we will end up with 2 less, or 98.

It takes a lot of thought and practice to come up with methods like this one. Your child may lack the imagination at first, or he may simply not be interested. Again, don't push him. As he acquires more experience, he will begin to choose the best methods automatically. Then, when he is really proficient at mental addition and subtraction, he is ready to start multiplication.

Multiplication is just a rapid form of addition. Instead of figuring out $8 + 8 + 8 + 8 + 8 + 8 + 8$ every time we need to know it, we just remember that seven eights are fifty-six. And once we know the simple combinations like this one, we can use them to work out the more difficult ones.

Many multiplication problems are too hard for the average person to solve mentally. There are methods for doing them, but most people find that it's not worth the trouble to learn them. If they need to know 3972×835, for example, they just sit down and work it out the way they were taught to in school.

There are many problems for which the tried and true methods are probably the fastest. But if you can add and subtract mentally, you can do a surprising number of multiplication problems that way, too. The link between these three processes is a property of numbers called the distributive law.

The distributive law has to do with the order in which we add and multiply. Suppose we were multiplying one number by the sum of several others, say 3 by $6 + 2 + 5 + 1$. Should we add 6, 2, 5 and 1 and then multiply the result by 3? Or should we multiply each of these numbers by 3 separately and then add the products?

According to the distributive law, it doesn't make any difference. We ought to get the same answer either way. $6 + 2 + 5 + 1 = 14$, and 14 times 3 is 42. On the other hand, 3×6 is 18, 3×2 is 6, 3×5 is 15 and 3×1 is 3. $18 + 6 + 15 + 3$ is also equal to 42. We can always "distribute" the product across the sum this way without changing the result.

The distributive law works just as well when there are minus signs involved. 3 times $6 + 2 - 5 - 1$ is the same as 3×6 plus 3×2 minus 3×5 minus 3×1. Try it and see.

Now, how can the distributive law help with mental multiplication? Take a problem like 6×87. Most people would write this one down right away. But with the help of the distributive law, it can be solved mentally in the time it takes to pick up a pencil.

The trick is to think of 87 as $80 + 7$. Then 6×87 is $6 \times (80 + 7)$, which is the same as $6 \times 80 + 6 \times 7$. Now there are two easy products to find instead of one hard one. 6×8 is 48, so 6×80 is just ten times as much, or 480. 6×7 is 42, and the answer to our problem is the sum of these: 522.

This use of the distributive law depends on the fact that some kinds of products are easier to find than others. Be sure your child knows the multiplication tables up to 9×9; if he doesn't, practice with a set of flashcards. Then show him how to multiply by ten; this just adds a zero to the number being multiplied. 7×10 is 70, 56×10 is 560 and 987×10 is 9870. This rule applies to multiples of ten as well; 7×6 is 42 so 7×60 is 420. And since 100 is 10 times 10, multiplying by 100 is the same as multiplying by ten twice, and this adds two zeros. In the same way, multiplying by 1000 adds three.

The method we used to multiply 6 by 87 can be used on any multiplication problem involving a two-digit number and a one-digit number. And it can be extended to many others. Try it on 6×807, 6×870 and 60×87. It applies to all of them, because they can all be changed to sums of products involving single digits or multiples of ten.

Some problems lend themselves better to differences of products. Take 76 × 98. At first glance it looks almost impossible to do this one mentally. 76 × (90 + 8) doesn't help much, and neither does 98 × (70 + 6). The trick is to realize that 98 can be expressed in terms of 100 and 2, both of which are very easy to handle. Think of 76 × 98 as 76 × (100 − 2). Then the answer becomes 7600 − 152, a far more workable expression than the others.

This technique can be applied to numbers near 10, 100, 1000, or any other power of ten. 9, for example, equals 10 − 1, so 57 × 9 equals 57 × (10 − 1), or 570 − 57. In the same way, 102 = 100 + 2, so 57 × 102 equals 57 × (100 + 2), which is 5700 + 114, or 5814.

Your child may find it awkward to do all this mentally, but even if he writes some of it down, this method should save him some time. Here is 4278 × 998 worked out in the usual way:

$$
\begin{array}{r}
4278 \\
998 \\
\hline
34224 \\
38502 \\
38502 \\
\hline
4269444
\end{array}
$$

There are a number of steps here, and the likelihood of an error is considerable.

The child who realizes that 998 = 1000 − 2 and knows how to apply the distributive law can do the same problem this way:

4278	4278000	(4278 × 1000)
× 2	− 8556	(4278 × 2)
8556	4269444	(4278 × 998)

All that's involved here is the doubling of 4278 and a simple subtraction.

If you've done a good job of introducing your child to mental arithmetic, he should be suggesting problems to you by this time. Perhaps he has already guessed that if it's easy to multiply by 1000 − 2, it should also be a simple matter to multiply by 1000 ÷ 2, or 500.

Products, like sums, obey commutative and associative laws, which means that numbers being multiplied together can be shuffled and grouped in any way at all. 43×500 is the same as $43 \times \dfrac{1000}{2}$, and this can be rewritten as $43 \times 1000 \times \dfrac{1}{2}$. The first two numbers can be multiplied simply by adding three zeros to the 43. All that remains is to multiply this result by $\dfrac{1}{2}$, and this is the same as dividing it by 2. So 43×500 is equal to half of 43000, or 21500.

This line of thinking opens up a whole new range of possibilities. 250 is half of 500, which is half of 1000; so to multiply any number by 250, just add three zeros and divide the result by 2 twice. (Most people find this easier than dividing by four.) Since 125 is half of 250, multiplication by 125 just involves one more division by 2. 5 is an easy number to work with too, because it is equal to $10 \div 2$. So to multiply by 5, just add zero and divide by 2. Similar rules can be formulated for numbers like 2500, $12\frac{1}{2}$, and 50, because all of them can be expressed as products involving fives and twos.

Many problems can be solved by combining these techniques. Take 472×35. 35 is $10 + 25$, and both of these numbers are easy to work with. To multiply by 10, just add zero. To multiply by 25, add two zeros and divide by 4. Then 472×35 becomes $4720 + \dfrac{47200}{4}$.

All that is required here is "short" division and addition. Most children are much faster at these than they are at multiplication, and more accurate, too.

The same methods can be applied to a host of other numbers. Show your child these examples, and then let him try to find some shortcuts of his own.

Here is a streamlined method for multiplying 673 by 15:

$$\begin{array}{r|l} 2 & 6730 \\ \hline & 3365 \\ \hline & 10095 \end{array}$$

The top number, 6730, is 673×10. Below it is half of it, or 673×5. And the sum of these two, in accordance with the distributive law, is 673×15.

125×2516 can be done in much the same way by taking advantage of the fact that 25 times a number is one-fourth of one hundred times it.

$$
\begin{array}{r|l}
4 \;|\underline{251600} & (2516 \times 100) \\
62900 & (2516 \times 25) \\
314500 & (2516 \times 125)
\end{array}
$$

And here is a quick way of multiplying 73 by 502.

$$
\begin{array}{r|l}
2 \;|\underline{73000} & (73 \times 1000) \\
36500 & (73 \times 500) \\
146 & (73 \times 2) \\
36646 & (73 \times 502)
\end{array}
$$

By now the conscientious parent must be asking a question he has raised many times before. Are these the right methods? They may be quick and easy, but won't it confuse the child to learn things one way at home and another at school?

The pursuit of the elusive "right" method deprives countless children of the help of parents who are well able to give it to them. Schools have no monopoly on logic and clear thinking, and there is no such thing as one "right" method. Some ways of getting an answer are longer than others, and some carry a higher risk of error, but any method that leads to a correct answer is "right." If the parent's thinking is a little different from the teacher's, so much the better. That gives the child two points of view instead of one.

The methods shown here for doing multiplication are not only right; they have a number of advantages over the conventional ones. To apply them intelligently, the child must use exactly the kind of imagination that the new math is designed to develop. He must understand the properties of numbers and adapt them to cover a wide range of possibilities. No good teacher would discourage this kind of thinking.

Division is related to multiplication in much the same way that subtraction is related to addition. 7×3 is 21, so $21 \div 3$ is 7 and $21 \div 7$ is 3. This means that many of the shortcuts for multiplication can be used for division too.

The distributive law can help us divide 315 by 3 mentally. Think of division by 3 as multiplication by $\frac{1}{3}$. And think of 315 as $300 + 15$. Then $\frac{315}{3}$ becomes $\frac{1}{3} \times (300 + 15)$, which is $\frac{300}{3} + \frac{15}{3}$. That equals $100 + 5$, or 105.

If the problem had been 231 divided by 3, it wouldn't have been so easy. $\frac{200}{3} + \frac{31}{3}$ is awkward because neither 200 nor 31 are evenly divisible by 3. $\frac{200}{3}$ equals $66\frac{2}{3}$ and $\frac{31}{3}$ equals $10\frac{1}{3}$. The sum of these two, 77, is the right answer, but this is not the best way to get it. It would have been much easier to think this way. The multiple of 3 just below 23 is 21. If 3 goes into 21, it goes into 210 too. So instead of writing 231 as $200 + 31$, write it as $210 + 21$. Then $\frac{231}{3} = \frac{210}{3} + \frac{21}{3}$, which gives 77 right away.

This method works even if there's a remainder. Take 193 divided by 3. 193 equals $180 + 13$, but 13 is not evenly divisible by 3. 12 is, though, so the problem becomes $\frac{1}{3}$ of $180 + 12 + 1$, which is $\frac{180}{3} + \frac{12}{3} + \frac{1}{3}$, or $64\frac{1}{3}$.

Dividing by 10 is especially easy. $\frac{193}{10}$ equals $\frac{190}{10} + \frac{3}{10}$, or 19.3. There's really no need to go to all this trouble, though. All we have to do to divide by 10 is move the decimal point one place to the left. 193 is understood to have a decimal point after the 3, so moving it one place to the left changes 193. into 19.3.

The rule for dividing by 100 is very similar. $\frac{193}{100}$ equals $\frac{100}{100} + \frac{93}{100}$, which is 1.93. Here the decimal point has been moved two places to the left. Division by 1000 would have moved it three places to give an answer of .193. And if we had wanted to divide 193 by 10,000, we would have moved the decimal point four places to the left, supplying a zero for the fourth place. $\dfrac{193}{10,000} = .0193$.

If it's easy to divide by 100, it should also be easy to divide by half of 100, or 50. The answer would be just twice as large. So the rule for dividing by 50 is to move the decimal point two places to the left and double the result. 3700 divided by 50 is 37×2, or 74. Similar rules can be formulated for 125, 250, and a host of others.

This is all well and good, but it still doesn't tell us what to do with a problem like $56\,\overline{\smash{)}9632}$. Of all the operations of arithmetic, long division is certainly the hardest. Most of us would do it this way:

$$
\begin{array}{r}
172 \\
56\,\overline{\smash{)}9632} \\
56 \\
\overline{403} \\
392 \\
\overline{112} \\
112 \\
\overline{0} \\
\end{array}
$$

The only hope of simplying this procedure is that old trick of turning a hard problem into an easy one. 56, remember, is 7×8. This makes 7 and 8 *factors* of 56. Dividing by the factors of a number instead of by the number itself can often make division easier. In this case, if we divide 9632 first by 7 and then by 8, we can get the answer by doing two short divisions instead of one long one. Then the work would look like this:

$$
\begin{array}{r}
7\,\overline{\smash{|}\,9632} \\
8\,\overline{\smash{|}\,1376} \\
\overline{172} \\
\end{array}
$$

Long division is hard because it's largely a process of trial and error. 56 is a number we know very little about. To divide it into 9632, we must decide how many times it goes into 96 and how many times it goes into 403. This takes a little thought. And if we guess wrong, we have to go back and do it again. 7 and 8, on the other hand, are right out of the multiplication table. We can divide both of them into any number at all, quickly and easily. Even if we find it difficult to do this mentally, dividing by 7 and 8 instead of by 56 will certainly shorten the written work.

Not all numbers are as easy to factor as 56 is. Luckily, there are simple divisibility tests for many of the smaller numbers, and it's a good idea to learn them.

Everyone knows the test for divisibility by 2. If a number ends in 0, 2, 4, 6, or 8, then 2 goes into it evenly. Otherwise it doesn't.

The rule for 3 is just as simple. To test a number for divisibility by 3, add its digits. If their sum is divisible by 3, so is the number. 3 goes into 7692, for instance, because $7 + 6 + 9 + 2$ equals 24. It does *not* go into 1907 because $1 + 9 + 0 + 7$ equals 17, which is not a multiple of 3.

If a number is divisible by 4, it is divisible by 2 twice. So if half a number is even, that number is a multiple of 4. 156 divided by 2 is 78, an even number, so 156 is divisible by 4. This is a reliable test, but not always a convenient one. Taking half of 1,734,016, for example, might take longer than we'd like it to. And fortunately, it isn't necessary.

If you list the multiples of 4 from 4 to 100 in one column and from 104 to 200 in another, you'll see that the last two digits repeat themselves. This is also true between 204 and 300, 304 and 400, and on as far as you care to go. If 4 goes into 16, it also goes into 116, 216, 716 and even 1,734,016. So no matter how long a number is, its last two digits are the key to divisibility by 4. If 4 goes into the number they form, it goes into the whole number.

5 goes into all numbers that end in 0 or 5, and no others.

Numbers that are divisible by 6 are divisible by both 2 and 3. So any *even* number whose digit sum is a multiple of 3 is evenly divisible by 6.

7 is a problem. There are a number of tests for divisibility by 7, but most of them take longer than regular division would. One way to see whether 7 is a factor of a number is to take away its final digit, double it, and subtract it from what's left of the number.

If 413 were being tested, for instance, the 3 would be removed, leaving a 41. 3 doubled equals 6, 41 − 6 is 35, and since 35 is divisible by 7, 413 is too. For larger numbers, the process has to be repeated.

The tests for 8 are similar to the ones for 4. 8 equals $2 \times 2 \times 2$, so if a number can be divided by 2 three times it can be divided by 8 once. Or you can just look at the last three digits. If the number they form is divisible by 8, the larger number is too.

9 follows exactly the same rule 3 does. If a number's digit sum is divisible by 9, so is the number.

10 is the easiest of all. It goes evenly into any number that ends in 0.

Applying these rules is usually very simple, but sometimes we hit a snag. Take 401. It isn't divisible by 2 or by any even number. It doesn't pass the tests for 3, 5 or 7 either, so there's no point in testing any multiples of these numbers. That rules out 9, 15 and 21. 11 doesn't go into it, and neither do 13, 17, or 19. It looks very much as though 401 has no factors at all.

But 401 is such an unfamiliar number. How about 23 or 37? Shouldn't we try those? The answer is no, and the reason is very simple. The best way to understand it is to look at a familiar number like 36.

36 has a great many factors. 2, 3, 4, 6, 9, 12, and 18 all go into it evenly. Here is a list of all the pairs of whole numbers whose products are 36.

$$1 \times 36$$
$$2 \times 18$$
$$3 \times 12$$
$$4 \times 9$$
$$6 \times 6$$
$$9 \times 4$$
$$12 \times 3$$
$$18 \times 2$$
$$36 \times 1$$

These numbers form an interesting pattern. Each pair of factors below 6×6 is duplicated, in reverse order, by a pair above it. Since the order of the factors makes no difference, the list would have been complete if we had stopped at 6×6.

6 is called the square root of 36 because it equals 36 when it is multiplied by itself, or squared. (There's a good reason for the use of the word "squared" here. A square whose area is 36 square inches has a side of 6 inches.) Every positive number has a square root, although most square roots aren't even. And it is never necessary to go above the square root of a number in looking for its factors. If we do, we only duplicate factors we already have.

What is the square root of 401? 20 × 20 is 400, and 21 × 21 is 441. Since 401 falls between 400 and 441, its square root falls between 20 and 21. We've tested all the numbers up to 21, and there's no need to go any further. 23 cannot be a factor of 401, and neither can 37. 401 has no factors but itself and 1, and this makes it a prime number.

So far we've considered the operations of arithmetic one at a time, but in everyday life they're much more likely to be used in combination.

A man is at 84th Street now, and he has to go down to 19th. How many miles will he travel if there are twenty city blocks to a mile? And what will the taxi fare be if he pays 45¢ for the first quarter of a mile and 10¢ for every quarter of a mile after that?

The British pound is worth $2.40. If the plane fare from London to New York is $480, how many pounds would it cost a family of two adults and three children to make the trip if each child travels for half fare?

One brand of tomato soup costs 14¢ for a 12 oz. can. Another sells at 2 cans for 25¢, but each contains only 10½ oz. Which brand is less expensive?

These are typical of the many problems that confront all of us in our everyday lives. Solving them will give your child a chance to practice his arithmetic and to see how very useful it can be. But it would be a mistake to give him the impression that numbers are nothing but tools.

Besides having utilitarian value, numbers are an endless source of fascination and amusement. They always have been. From time immemorial, people have enjoyed discovering the patterns they make and the reasons they behave the way they do. And children, even very young ones, can enjoy making these discoveries too.

II

The Shapes of Numbers

GIVE YOUR CHILD twenty pennies and ask him if he can make four different sized triangles with them. Each triangle must have the same number of pennies on all three sides. He should have no trouble forming triangles with 3, 6, and 10 pennies, and perhaps he will even realize that the remaining penny forms a triangle of its own.

These four numbers, 1, 3, 6, and 10, are the first four triangular numbers.

9 is a square number.

And 5 is pentagonal.

```
            O

        O       O

          O   O
```

Numbers that have shapes associated with them are called figurate numbers. Children understand concrete things best, and figurate numbers have great appeal for them. They enjoy forming them, building one upon another, and exploring the relationships between them.

Square numbers are probably the best ones to start with. The first of these is 1, because a perfect square can be drawn around a single penny. Show your child how this is done and ask him to see if he can form the next square number. He will probably realize that three more pennies are needed to do this, so the next square, with 2 pennies on each side, will contain $1 + 3$ or 4 pennies. Viewed another way, it has 2 rows of pennies with 2 pennies in each row.

```
    O  O              O  O
    ‾‾‾
    O| O              O  O
  1 + 3              2 × 2
```

To build the next square number on this one, 5 pennies must be added, 2 on top, 2 on one side, and 1 in the corner between them. The resulting square contains 9 pennies, 3 rows of 3 each.

```
   O  O  O           O  O  O
   ‾‾‾‾‾
   O  O| O           O  O  O
   ‾‾‾
   O| O| O           O  O  O
  1 + 3 + 5           3 × 3
```

And the next square requires 7 more pennies to give a total of 4×4, or 16.

```
  O  O  O  O         O  O  O  O
  ‾‾‾‾‾‾‾‾
  O  O  O| O         O  O  O  O
  ‾‾‾‾‾
  O  O| O| O         O  O  O  O
  ‾‾‾
  O| O| O| O         O  O  O  O
  1 + 3 + 5 + 7        4 × 4
```

These results are especially interesting if they're listed this way:

$$1 + 3 = 2 \times 2$$
$$1 + 3 + 5 = 3 \times 3$$
$$1 + 3 + 5 + 7 = 4 \times 4$$

A pattern seems to be emerging here, and the child who is properly guided can discover it for himself. The first 2 odd numbers added together equal 2×2. The first 3 equal 3×3. And the first 4 equal 4×4. Does this pattern continue? Is the sum of the first 5 odd numbers 5×5? Do the first 11 equal 11×11? How about the first 17?

This kind of situation is a mathematician's delight. Once he recognizes and understands it, he can do a whole family of problems with very little effort. What is the sum of all the odd numbers from 1 to 207? We should be able to find out by doing a simple multiplication problem, but what numbers shall we multiply?

To find out, we have to investigate the pattern more closely.

1 is the 1st odd number.
3 is the 2nd odd number.
5 is the 3rd odd number.
7 is the 4th odd number.
9 is the 5th odd number.

How are the numbers on the left related to the ones on the right? Well, half of $9 + 1$ is 5. Half of $7 + 1$ is 4. And half of $5 + 1$ is 3. Adding one to each number on the left and dividing the result in half gives the number opposite it on the right. Since half of $207 + 1$ is 104, 207 must be the 104th odd number. So the sum of all the odd numbers from 1 to 207 is 104×104, or 10,816.

This is a very useful rule, and it would be a good idea to state it as clearly as possible. To add all the odd numbers, starting with 1 and ending with any number you choose, add 1 to your chosen number, divide the result by 2, and multiply the number you get by itself.

There's an even better way of expressing this. Let N stand for the chosen number. Then $\dfrac{N+1}{2}$ is the number you multiply by itself to add all the odd numbers from 1 to N.

Triangular numbers reveal a similar pattern. The first of these is 1.

○

The second is formed by adding 2 more pennies.

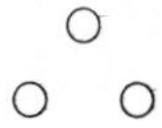

○
○ ○

The next requires 3 more

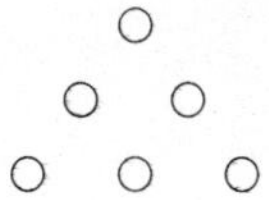

○
○ ○
○ ○ ○

And the next 4.

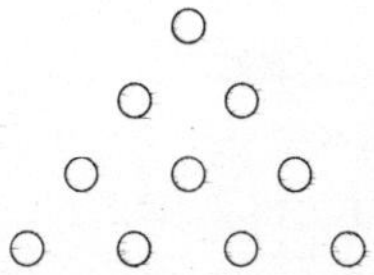

○
○ ○
○ ○ ○
○ ○ ○ ○

If your child can recognize the pattern here, he should be able to list the first ten triangular numbers without using any more pennies. Here they are:

1	= 1
1 + 2	= 3
1 + 2 + 3	= 6
1 + 2 + 3 + 4	= 10
1 + 2 + 3 + 4 + 5	= 15
1 + 2 + 3 + 4 + 5 + 6	= 21
1 + 2 + 3 + 4 + 5 + 6 + 7	= 28
1 + 2 + 3 + 4 + 5 + 6 + 7 + 8	= 36
1 + 2 + 3 + 4 + 5 + 6 + 7 + 8 + 9	= 45
1 + 2 + 3 + 4 + 5 + 6 + 7 + 8 + 9 + 10	= 55

And here are the first ten square numbers listed next to them:

1	1
4	3
9	6
16	10
25	15
36	21
49	28
64	36
81	45
100	55

Look at these two columns, and see if you can find a relationship between them. If you can't, perhaps your child will discover it for you. Every square number is equal to the sum of two triangular ones, the number just opposite it and the one just above. 4 equals 3 + 1. 9 equals 6 + 3. And moving down a little further, 64 equals 36 + 28. 81 equals 45 + 36.

Go back to the pennies again, and you'll see why this is so. Take the square with 6 pennies on each side. If you draw a line through it, you can turn it into two triangles, one with 6 pennies on a side and one with 5.

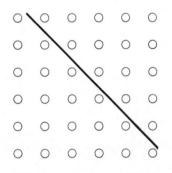

Every square can be divided into two triangles this way, one with as many pennies on a side as the square has, the other with one less.

This pattern may also make for easy problem solving. The 25th triangular number is the sum of all the whole numbers from 1 to 25. Adding these one by one would be a very tedious job. If we could find an easy way to get triangular numbers, this same addition could be accomplished automatically.

We know that the 25th square number is the sum of the 24th and 25th triangular numbers. The problem is in knowing just how to split it up to get them. To find out, we must go back to the pennies again.

We have already seen how a square of 36 pennies can be divided into two triangles. One of these contains 6 pennies more than the other. Suppose we were to draw a second line, cutting off these 6 extra pennies. Then we would have two identical triangles with 5 pennies on a side and a row of 6 pennies between them.

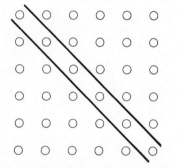

The number of pennies in both triangles together can be found by taking 6 away from 36, and each triangle contains just half this amount. So the 5th triangular number must be half of $6 \times 6 - 6$.

Is this a pattern we can count on? Does the 7th triangular number equal half of $8 \times 8 - 8$? Does the 9th equal half of $10 \times 10 - 10$? Try them and see.

Once your child is convinced that this pattern continues to hold, he should be able to add all the numbers from 1 to 100 in less than a minute by finding the 100th triangular number. All he has to do is multiply 101 by 101, subtract 101 from this product, and divide the answer in half. What a wonderful opportunity to practice mental arithmetic!

It's really remarkable, when you think about it, to discover such a simple way of adding a hundred numbers just by looking at arrangements of pennies on a table. And it's just as easy to add a thousand. This is the kind of mathematics that captures the imagination of a child and makes him eager for more.

III

Magic Squares

HERE'S A GAME you can play with any child who knows the addition facts from 1 to 9. Use nine playing cards and arrange them this way.

$$
\begin{array}{ccc}
8 & 1 & 6 \\
3 & 5 & 7 \\
4 & 9 & 2
\end{array}
$$

The players take turns choosing cards, and the winner is the first one to get three whose sum is 15. There are a number of ways to do this. 8, 1, and 6 is one winning combination, 4, 9, and 2 is another. If you look closely, you'll see that there are six more, and that all of them fall along rows, columns, or diagonals. As a matter of fact, this is just a glorified version of tick-tack-toe. If you have a winning strategy for that game, you can use it for this one too.

Any square of numbers in which the rows, columns, and diagonals have the same sum is called a magic square. This one originated in ancient China, where it was called the lo-shu. It is a third order square because there are three numbers on each side of it.

Magic squares can be rearranged in a number of ways without losing their magic. In the case of the lo-shu, the outside columns can be interchanged,

```
6  1  8
7  5  3
2  9  4
```

and so can the top and bottom rows.

```
4  9  2
3  5  7
8  1  6
```

The rows can also be turned into columns.

```
8  3  4
1  5  9
6  7  2
```

There are eight different rearrangements altogether, and you and your child should be able to find all of them.

The lo-shu is the only third order magic square, but for the fourth order there are 880. No one knows exactly how many fifth order magic squares there are, but the number is probably over 13 million!

Magic squares have been a source of fascination for centuries, and children of all ages enjoy making them. Any odd order square can be constructed with a simple set of rules. Let's see how they work for order seven. We start by putting down 49 dots to guide us and placing a 1 at the top of the middle column.

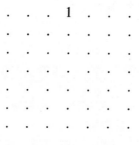

Whenever possible, the numbers should follow one another along diagonals that move upward to the right, but we can't put 2 to the right above 1, because 1 is already at the top of the square. So we follow *Rule I: Any number that comes after a number in the top row of the square is written at the bottom of the next column to the right.* The only exception to this is the number in the upper right hand corner; its successor goes directly below it.

Having followed Rule I, we can place 3 and 4 along a diagonal, but with 4 we come to the last column of the square.

```
    .   .   .   1   .   .   .

    .   .   .   .   .   .   .

    .   .   .   .   .   .   .

    .   .   .   .   .   .   .

    .   .   .   .   .   .   4

    .   .   .   .   .   3   .

    .   .   .   .   2   .   .
```

Now we need *Rule II: Any number that comes after a number in the last column on the right* (except the one in the upper right hand corner) *is written at the left end of the row directly above it.* 5, 6, and 7 take their places accordingly, but the space that 8 should occupy is already taken up by 1.

```
    .   .   .   1   .   .   .

    .   .   7   .   .   .   .

    .   6   .   .   .   .   .

    5   .   .   .   .   .   .

    .   .   .   .   .   .   4

    .   .   .   .   .   3   .

    .   .   .   .   2   .   .
```

This situation calls for *Rule III: When the space a number should occupy is already taken, the number goes directly below its predecessor.*

From here on it's smooth sailing. 8 goes below 7; 9 and 10 are on a diagonal, and since 10 is at the top of the square we follow Rule I and put 11 under 3. Here is the square with some of the key numbers filled in. See if your child can complete it himself before looking at the finished square below it.

.	39	.	1	10	.	.
.	.	7	9	.	.	29
.	6	8	.	.	.	.
5	.	.	.	.	36	.
.	15	.	.	.	.	4
21	.	.	.	43	3	.
.	.	.	49	2	11	.

30	39	48	1	10	19	28
38	47	7	9	18	27	29
46	6	8	17	26	35	37
5	14	16	25	34	36	45
13	15	24	33	42	44	4
21	23	32	41	43	3	12
22	31	40	49	2	11	20

If you check on the square's "magic" by adding its rows, you'll find that each of them totals 175. This number is associated with all seventh order squares and is called the seventh order constant. Is there any way we could have predicted it? How is 175 related to 7?

The seventh order square contains all the whole numbers from 1 to 49. Since the sum of these is the 49th triangular number, all seven rows together add up to half of $50 \times 50 - 50$, or 1225. And since all the rows have the same sum, a single row must equal 1/7 of 1225, or 175.

If the seventh order constant is 1/7 of the 49th triangular number, the third order constant must be 1/3 of the 9th triangular number. We've already added the rows of the lo-shu and found their sums to be 15. Let's check this with the formula. The 9th triangular number is half of $10 \times 10 - 10$, or 45. And 1/3 of 45, to be sure, is 15.

It's easy to construct a fifth order magic square and show that each of its rows adds up to 1/5 of the 25th triangular number. If your child is really interested, he should compute the ninth and eleventh order constants and check them by actually constructing the squares. This requires a lot of patience; one mistake can throw everything off. But there are certain built-in-checks.

If you look at the seventh order square, you will see that one diagonal is an unbroken series of numbers. Not only that, but 25, the number which is just halfway between 1 and 49, lies at the very center of the square. The same thing is true of the lo-shu; 4, 5 and 6 fall on a diagonal with 5 in the central position. All magic squares constructed by this method have this same property. In the ninth order square, for example, 41 should be at the center, because 41 is just halfway between 1 and 81. And all the numbers from 37 to 45 should fall on an unbroken diagonal. If they don't, there has been a mistake. When the rules are followed properly, every space in the square is filled automatically, with the last number at the bottom of the middle row.

Here are three incomplete squares to serve as guides if you need them.

Order Five

	24			
				16
4	6			
			21	
11	18		2	9

Order Nine

```
 .   .  69   .   .   .   .   .  45
 .   .   .   .   .   .   .   .   .
 .   .   .  10   .   .   .   .   .
77   .   .   .   .   .   .  55   .
 6   .  19   .   .   .   .   .   .
 .   .   .   .   .  64   .   .   .
26  28   .   .   .   .   .   .   .
 .   .   .  60   .  73   .   .   .
37  48   .   .   .   .  13  24  35
```

Order Eleven

```
68   .  94   .   .   .   .   .  40   .  66
80   .   .   .   .   .   .   .   .   .   .
 .   .   .   .  12   .   .   .   .   .   .
 .   .   .   .   .   .   .   .   .  78   .
 .   .   .  23   .   .   .   .   .   .   .
 7   .   .   .   .   .   .   .  89   .   .
 .   .  34   .   .   .   .   .   .   .   .
 .   .   .   .   .   .   . 100   .   .  18
43   .   .   .   .   .   .   .   .   .   .
55   .   .   .   . 111   .   .   .   .   .
 .  69   .   . 108   .   .   .  28   .   .
```

There's a very simple way to make a fourth order magic square. First write the numbers from 1 to 16 in four rows of four.

$$1 \quad 2 \quad 3 \quad 4$$
$$5 \quad 6 \quad 7 \quad 8$$
$$9 \quad 10 \quad 11 \quad 12$$
$$13 \quad 14 \quad 15 \quad 16$$

Then remove the diagonals,

$$\begin{array}{cccc} & 2 & 3 & \\ 5 & & & 8 \\ 9 & & & 12 \\ & 14 & 15 & \end{array}$$

and put them back upside down, so that the lower right hand corner becomes the upper left hand one.

$$\begin{array}{cccc} 16 & 2 & 3 & 13 \\ 5 & 11 & 10 & 8 \\ 9 & 7 & 6 & 12 \\ 4 & 14 & 15 & 1 \end{array}$$

The square that results is far more magic than any you've seen so far. Like all magic squares, this one's rows, columns, and diagonals all have the same sum. For the fourth order the constant is 34, 1/4 of the 16th triangular number. But in this square the rows, columns, and diagonals are only the beginning.

This is called a symmetric square because of the special way its numbers are arranged. To see the pattern, draw straight lines between every pair of numbers whose sum is 17. You'll find that all of them go through the center.

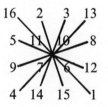

Within this fourth order square there are many small squares and rectangles whose corners add up to the magic constant. The diagrams below show you where they are.

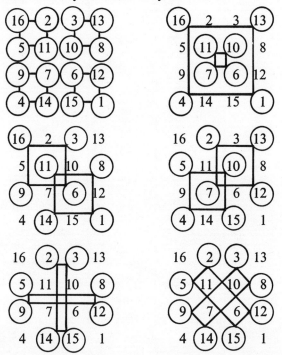

Here is a diabolic magic square, the most remarkable type of all.

$$
\begin{array}{cccc}
7 & 12 & 1 & 14 \\
2 & 13 & 8 & 11 \\
16 & 3 & 10 & 5 \\
9 & 6 & 15 & 4
\end{array}
$$

Diabolic squares have all the properties symmetric ones do, and more besides. This one contains four additional squares of 34.

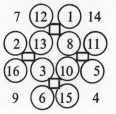

and three extra groups of 34 among its diagonals. The easiest way to see these is to put one diabolic square on top of another.

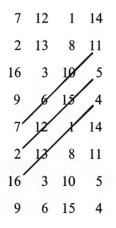

If you shift a column of a diabolic square from one end to the other, you get another diabolic square. The same applies to the rows; they can be moved from the top to the bottom or the bottom to the top. This means that if four diabolic squares are put together to form an array of numbers, every fourth order square in the array is diabolic.

7	12	1	14	7	12	1	14
2	13	8	11	2	13	8	11
16	3	10	5	16	3	10	5
9	6	15	4	9	6	15	4
7	12	1	14	7	12	1	14
2	13	8	11	2	13	8	11
16	3	10	5	16	3	10	5
9	6	15	4	9	6	15	4

How many magic squares can you find among these numbers?

Not all magic designs are squares. Here's a magic triangle.

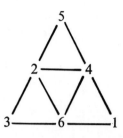

The numbers on each of its sides add up to 10, and the three numbers on each of the small corner triangles total 11.

And here's a magic cube.

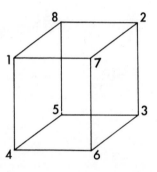

The corners of each of its faces have a sum of 18.

These are four circles in this picture, and the sum of the numbers on each of them is the same.

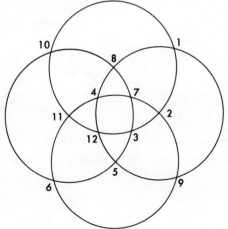

In this five-pointed star, the four numbers on any one line add up to 24.

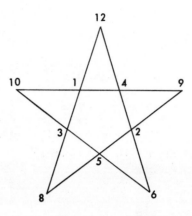

This magic hexagon is the only one of its kind.

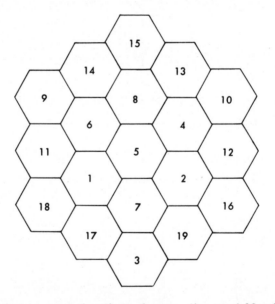

All the little hexagons that lie in the same line total 38, whether there are three, four, or five of them.

There's no limit to these magic shapes, and if your child stretches his mathematical imagination a little, he may enjoy inventing some of his own.

IV

Number Bases and the Game of Nim

ONE, TWO, THREE, four, five, six, seven, eight, nine, decim, decimun, ten. That's the way we might count if we had six fingers on each hand. What's that you're saying? We don't, so what's all the fuss about? Just this. A number system based on ten, as ours is, has many disadvantages. There are times when other systems of numeration are far more useful. Here's a fascinating little game that you can win easily if you learn another way of counting.

Arrange ten pennies in rows of four, three, two and one, like this.

Now you are ready to challenge your son or daughter to a game of Nim.

The rules are very simple. Each player takes his turn by removing any number of pennies from any one row. The game continues until all the pennies are gone. Whoever gets the last one wins.

If you play Nim often enough your game is bound to improve, but it won't be perfect unless you discover the one opening move that is a sure winner. Start with this winning move, continue to play correctly, and you will be absolutely unbeatable.

The variations of Nim are endless. It can be played with any number of pennies in a row and any number of rows. But no matter how complicated the arrangement is, there's always a simple way to win the game. The strategy is easy enough to be mastered by a child of eight, and anyone who knows it can beat anyone who doesn't, so if you want to be the Nim champion in your house, read the next few pages of this book and then hide it.

To win at Nim you must be able to write binary numbers. The binary number system is based on two in exactly the way that our decimal system is based on ten. Both are called positional number systems, because the value of each digit in a number depends on the position it occupies. 73 and 37, for example, mean two very different things. In the first case there are 7 tens and 3 ones, in the second case 3 tens and 7 ones.

The digit on the extreme right in any number is the ones digit. In our decimal system, the next digit to the left is the tens digit, followed by the hundreds digit and the thousands digit. When we write 5,437 we mean 5 thousands, 4 hundreds, 3 tens, and 7 ones. One hundred and one thousand are called powers of ten because they result from multiplying ten by itself. One hundred is ten times ten; one thousand is ten times ten times ten.

In the binary system, powers of two replace powers of ten. The digit on the extreme right is still the ones digit, but the next digit to the left becomes the twos digit. Next to that is the digit for two times two, or four, then the one for two times two times two, or eight, and so forth.

The only number that is the same in both the binary and decimal systems is one. To represent two in the binary system, we put a one in the twos place and a zero in the ones place, meaning that there is one two and no ones. Three is the sum of one two and one one, so it is written 11. To write four, we put a one in the fours place and follow it by two zeros (one four, no twos, and no ones). The first four binary numbers are therefore 1, 10, 11, and 100.

The binary numbers from one to fifteen are listed in the following table.

decimal number	eights	fours	twos	ones	binary number
1				1	1
2			1	0	10
3			1	1	11
4		1	0	0	100
5		1	0	1	101
6		1	1	0	110
7		1	1	1	111
8	1	0	0	0	1000
9	1	0	0	1	1001
10	1	0	1	0	1010
11	1	0	1	1	1011
12	1	1	0	0	1100
13	1	1	0	1	1101
14	1	1	1	0	1110
15	1	1	1	1	1111

Now you're ready to become an instant Nim champion. Before you take your first turn, write down the number of pennies in each row, using binary numbers. Then add the results in the usual way: as if they were ordinary decimal numbers. If all the digits in the sum are even, the position is a safe one for you, and your partner should go first. If any of the digits in the sum are odd, the position is unsafe, and you must make these odd digits even by adding or removing ones. Once you've done this, your opponent's move will automatically make the position unsafe again. Continue turning unsafe positions into safe ones, and you'll win the game.

Here are the binary numbers for the four, three, two, one configuration.

$$
\begin{array}{r}
100 \\
11 \\
10 \\
\underline{1} \\
122
\end{array}
$$

This position is unsafe because the sum contains a one. The only way to eliminate it without introducing any other odd numbers is to remove the top row completely, and this is the winning move referred to earlier. Confronted with the three, two, one configuration that remains, your opponent is licked before he starts.

Sometimes there's more than one way to make a position safe. Suppose you were faced with this one.

The binary numbers for these rows are 111 (7), 101 (5), 100 (4), and 11 (3). Their sum contains two threes.

$$
\begin{array}{r}
111 \\
101 \\
100 \\
11 \\
\hline
323
\end{array}
$$

There are two ways to change these threes to twos. One is to make the 111 a 10; the other is to remove the 101. 111 in the binary system is seven, 10 is two, so the first possibility means removing five pennies from the top row. The second means removing the next row altogether.

Here are some more problems for you to practice on.

In binary numbers these rows correspond to 10, 11, 10, and 1, which total 32. There are two ways to change this 32 to a 22; either remove a 10 or turn the 11 into a 1. This means taking away one of the rows containing two pennies or removing two of the pennies from the row containing three.

```
O   O   O   O   O   O   O
                    O   O
            O   O   O
                O
                O
```

Here the rows correspond to 111, 10, 11, 1, and 1. Their sum is 134, and the only move that will make both the 1 and the 3 even is taking away six pennies from the row containing seven. This turns the 111 into a 1 so that the sum becomes 24.

```
    O   O   O   O   O   O
    O   O   O   O   O   O
        O   O   O   O   O
            O   O   O   O
                    O
```

This position is a safe one; the rows are 110, 110, 101, 100, and 1, and their total is 422. Try to get your partner to go first, but don't worry too much if he refuses. He's almost certain to make a mistake eventually, and as soon as he does your victory is assured.

```
O   O   O   O   O   O   O   O   O   O
                        O   O   O
                O   O   O   O
```

Here the sum of 1010, 11, and 100 is 1121. To make it 222, the 1010 must be changed to a 111, and this means removing three pennies from the row containing ten.

```
O   O   O   O   O   O   O   O   O   O   O   O   O   O
        O   O   O   O   O   O   O   O   O   O
            O   O   O   O   O   O   O   O
                    O   O   O   O   O
                        O   O   O   O
                            O   O   O
```

There are three possibilities here:

a. Remove the row of fourteen completely.

b. Change the 1010 to a 100, which means removing six pennies from the row containing ten.

c. Change the 1000 to a 110, which means removing two pennies from the row containing eight.

When your child finds it impossible to beat you at Nim, he'll be eager to learn the winning system himself, and the knowledge of binary numbers that he acquires in the process may well prove useful to him. Once a mathematical curiosity, the binary system now forms the basis for the operation of digital computers. Its tremendous importance has led many schools to include it in the modern math curriculum.

The binary system is invaluable to computer designers because it has only two digits, 1 and 0. Switches are either on or off, and electric charges are positive or negative. Any such pair of opposing signals can be combined to express numbers in binary form.

If children were taught the binary system in school, they would have much longer numbers to write, but they wouldn't have to memorize multiplication tables. There are only four simple multiplication facts in the binary system:

$$1 \times 1 = 1$$
$$1 \times 0 = 0$$
$$0 \times 1 = 0$$
$$0 \times 0 = 0$$

Binary addition is easy too:

$$1 + 1 = 10$$
$$1 + 1 + 1 = 11$$
$$1 + 1 + 1 + 1 = 100, \text{ and so forth.}$$

This is the way a computer would multiply 26 by 7:

$$
\begin{array}{r}
11010 \\
111 \\
\hline
11010 \\
11010 \\
11010 \\
\hline
10110110
\end{array}
$$

The next time your child has addition problems to do for homework, he may enjoy checking them with binary arithmetic.

Although ten and two are used most commonly, any number can serve as the base of a positional number system. In the ternary system (base three) there are three digits, 0, 1, and 2. The place to the left of the ones digit is the threes digit, next to that is the digit for three times three, or nine, and so on. The ternary numbers from one to twenty-six are listed in the table below.

decimal number	nines	threes	ones	ternary number
1			1	1
2			2	2
3		1	0	10
4		1	1	11
5		1	2	12
6		2	0	20
7		2	1	21
8		2	2	22
9	1	0	0	100
10	1	0	1	101
11	1	0	2	102
12	1	1	0	110
13	1	1	1	111
14	1	1	2	112
15	1	2	0	120
16	1	2	1	121
17	1	2	2	122
18	2	0	0	200
19	2	0	1	201
20	2	0	2	202
21	2	1	0	210
22	2	1	1	211
23	2	1	2	212
24	2	2	0	220
25	2	2	1	221
26	2	2	2	222

Ternary numbers form the basis for a very ingenious card trick. Ask your child to select one of twenty-seven cards without telling you which it is. Then let him deal the cards from left to right, face side up, into three piles of nine and tell you which pile contains his chosen card. He may then put the piles together, still face side up, in any order he wishes. Watch carefully to see where he puts the pile containing the chosen card, and write down a 0, 1, or 2 according to whether this pile is on the top, in the middle, or on the bottom. Then have him turn the cards over and repeat the process two more times. Put each digit you record to the left of its predecessor.

Taken together, the three digits you have written will be one of the twenty-six ternary numbers in the chart, and this number tells you how many cards lie between the chosen card and the top of the pack when it is held face side up. Suppose the pile containing the chosen card was put on the top the first time, on the bottom the second, and in the middle the third. Your ternary number would be 120, or fifteen. Hold the pack face up, count off fifteen cards, and the next will be the one your child chose.

In principle, this trick can be done with any number base, but three is by far the most practical. In base four, for instance, four piles of cards would be needed with sixty-four cards in each, and these 256 cards would have to be dealt four different times. Base five would involve 3125 cards! In the binary system, on the other hand, only four cards would be used, not enough to be interesting.

Choosing a number system is largely a matter of convenience. The binary system is best for playing Nim; the ternary is best for this particular card trick. All these systems work as well as our decimal system does, and some work even better. Our use of ten as a base is just a biological accident, and many mathematicians think it's not a very happy one.

Twelve would be a better base than ten, because it's divisible by three as well as by two. In base ten the multiples of three can end in any digit at all. In base twelve they all end in 0, 3, 6, or 9, and so do the multiples of nine. The multiples of four and eight follow a similar pattern, and the multiples of six do, too. But in exchange for this convenience, base twelve requires two new digits and a larger multiplication table.

Base six has many of the advantages of base twelve without any of the disadvantages. And as an added bonus, in base six all the prime numbers except two and three end in 1 or 5.

Of course, changing to another base would mean learning new tables for multiplication and addition. Here they are for base six.

addition							*multiplication*					
+	1	2	3	4	5	10	×	1	2	3	4	5	10
1	2	3	4	5	10	11	1	1	2	3	4	5	10
2	3	4	5	10	11	12	2	2	4	10	12	14	20
3	4	5	10	11	12	13	3	3	10	13	20	23	30
4	5	10	11	12	13	14	4	4	12	20	24	32	40
5	10	11	12	13	14	15	5	5	14	23	32	41	50
10	11	12	13	14	15	20	10	10	20	30	40	50	100

The rules of arithmetic work in any base, and children enjoy the novelty of multiplying with a system in which $5 + 4 = 13$ and $2 \times 3 = 10$. It's a good experience for them, too, because it enhances their understanding of their own number system and the rules that govern it.

V

Topology

WHEN IS A TEACUP like a doughnut? This may sound like a riddle, but it really isn't. If a teacup were made out of soft clay, it could be stretched and rolled until it was shaped exactly the way a doughnut is.

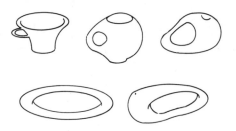

The space between the cup and its ear would become the hole in the doughnut; so no connections would have to be made or broken. For this reason, the cup and the doughnut are said to be *topologically equivalent*.

Topology is one of the newest branches of mathematics and one of the hardest to define. Probably the best way to get a feeling for what topology is all about is to study some of the problems it deals with.

One of the very first problems in topology originated more than two hundred years ago in the town of Königsberg in East Prussia. Through Königsberg ran the Pregel River, and in the river there were two islands. One bridge connected the islands to each other; six connected the islands to the shores.

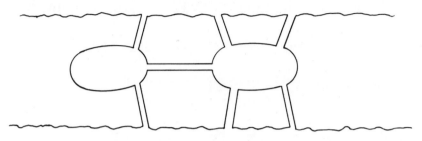

The people of Königsberg liked to stroll, and they particularly enjoyed strolling across the bridges. While they walked, they tried to answer this question: was it possible to plan a route which would cross each of the seven bridges once and only once?

Make a sketch of the bridges for your child, and see if he can discover such a route. Don't be disappointed if he can't, though; the people of Königsberg couldn't either. After a while, in fact, they began to suspect that there wasn't any.

It was the great Swiss mathematician Leonhard Euler who finally proved that they were right. First he shrank the bridges to lines and the islands to points.

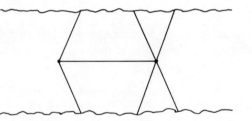

Then he shrank the shores to points too.

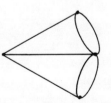

If he could trace every line in the figure without taking his pencil off the paper, then he could cross all the Königsberg bridges as well.

Euler studied this network and other networks like it. In some, like the one below, it was possible to start anywhere, go over every line exactly once, and come out at the starting point again.

In others, like this one, each path could be traced once if the right starting point were chosen, but the path would not end where it began.

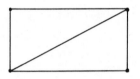

In still others, like the Königsberg bridges, it seemed to be impossible to trace the whole figure without going over some of the lines twice.

Euler was able to prove that these networks are governed by a simple set of rules. The rules have to do with the number of lines that are connected to each of the points. Euler called the points *nodes*. If an odd number of lines is connected to a point, it is called an odd node; otherwise it is an even node.

A network having no odd nodes can always be traced completely, and any of its nodes can serve as a starting point. In diagram A above, there are four lines connected to each of the two points on the circle, four to the point at the center, and two to each of the others. Each point is an even node, so no matter where you begin you can trace the entire figure and return to your starting point.

Only one other type of network can be traced completely, and that is one that has just two odd nodes. In diagram B, two of the nodes are joined to two lines and two to three. If you start at one of the even nodes, you'll come back to it without having traced all the lines in the network. The only way to trace the network completely is to start at one of the odd nodes. When you do, you'll automatically end up at the other one.

Networks with only one odd node, or with three or more, can never be traced completely. In the Königsberg bridge network, all four nodes are odd. Three of them have three lines attached to them, and the fourth has five. It is impossible to trace the entire network without going over some of the lines twice. (An eighth bridge has been added since Euler's time, which connects the two banks directly, like this.

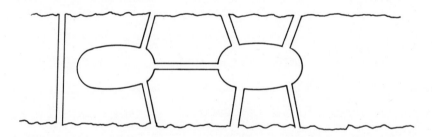

Does this new bridge solve the problem?)

From this study of the Königsberg bridges, Euler deduced a set of rules that apply to all networks, no matter how complicated they are. By counting the lines at each node, it is always possible to decide whether a network is traceable or not. And it doesn't have to be in two dimensions either. The edges of this box can't be traced because three of them meet at each corner.

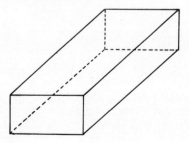

Euler made another discovery about solids like this box. This one had to do with faces, edges, and vertices. The faces of the box are the six rectangles that form its top, bottom, and sides. The edges are the lines in which the faces meet. And the vertices are the points that we call corners.

This box has six faces, twelve edges, and eight vertices; if we add the number of faces to the number of vertices and subtract the number of edges, we get six plus eight minus twelve, or two. There is nothing particularly remarkable about this. What *is* remarkable is Euler's discovery that this formula, faces plus vertices minus edges, gives two *for any solid of this type.*

The solids to which this formula applies are called *polyhedra.* Polyhedra are bounded by flat planes that meet in straight lines which in turn meet in points. Houses are polyhedra.

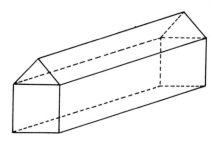

So are pyramids.

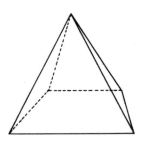

In the house pictured above, there are nine faces, seventeen edges, and ten vertices. Nine plus ten minus seventeen is again two. See if you can get the same result for the pyramid.

There are five regular polyhedra, in which all the edges are equal and all the faces have the same shape and size. The first is the regular *tetrahedron*; it has four faces, and each is an equilateral triangle. You can make it from the pattern below. Cut on the solid lines, fold on the dotted ones, and attach the edges with transparent tape.

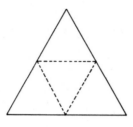

Two of the other regular polyhedra are also formed from equilateral triangles. The *octahedron* is made from eight of them, the *icosahedron* from twenty.

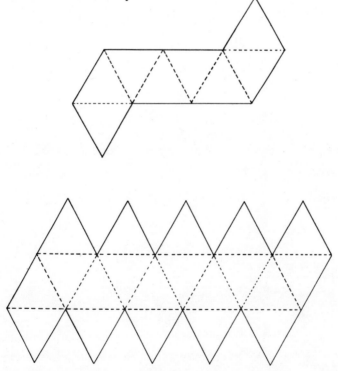

The most common regular polyhedron is the *cube*; it is made from six squares.

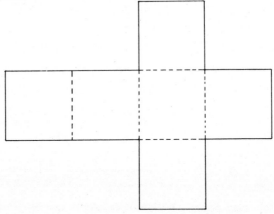

And the last of the five, the *dodecahedron*, is made from twelve pentagons.

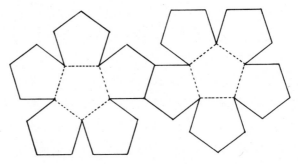

When you're finished, the five solids should look like this.

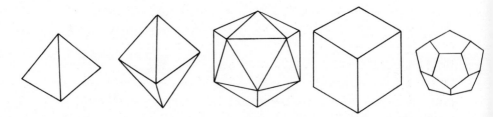

All children enjoy constructing these solids, and they offer beautiful confirmation of Euler's formula. For each of them, different as they are, faces plus vertices minus edges equals two. This two is called a *topological invariant*, because it is the same for all polyhedra.

Here's a topological trick that can provide entertainment at your next children's party. Before the guests arrive, make some long, narrow paper loops from adding machine tape. In every loop but one, put the ends together in the ordinary way.

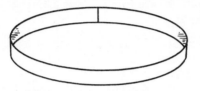

In the last loop, give the tape a half twist before connecting the edges.

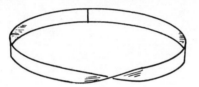

No one will notice it if the strips are long enough.

Give each child a loop and a pair of scissors, and see who can split his loop in two the fastest. The way to do this, of course, is to cut it right down the middle. For the untwisted loops, this procedure is perfectly straightforward. But the child whose loop has the half twist is in for a surprise. When he's through cutting he'll still have a single loop, but it will be twice as long as the one he started with, and it will have four half twists in it instead of one.

The twisted loop is called a *Moebius strip*, and it has a number of fascinating properties. An ordinary loop has two edges. An ant could walk all the way around one of them without ever touching the other. But if the ant walked around the edge of this Moebius strip, he would cover every bit of it before he returned to his starting point. That's because the Moebius strip has only one edge. Make a small one from a long narrow strip of paper (adding machine tape is fine) and trace the edge with your finger.

If you draw a line down the middle of an ordinary loop until it meets itself, one side will have the line on it and the other side won't. Now try the same thing with the Moebius strip. Before the line meets itself it will have covered every bit of the surface, because the Moebius strip has only one side.

One edge and one side; these account for the Moebius strip's astonishing properties. If you cut it along its length $\frac{1}{3}$ of the way from its edge, you'll get another surprising result. This time there are two loops, and they're linked! One has a single half twist in it, and the other has two.

To liven up your party a bit, you might try giving some of the loops two or more half twists. With two, the cutting will produce a pair of linked loops with two half twists in each.

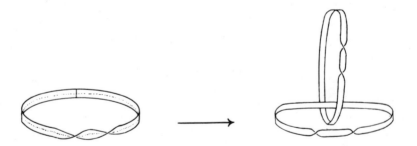

Start with three half twists, and you'll have the biggest surprise yet. This time you'll get a single strip, but there'll be a knot in it.

The double Moebius strip is interesting too. Cut two identical strips and place one on top of the other. Now, holding them together, make a single half twist and connect both pairs of ends.

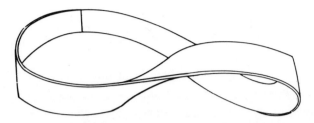

If you run your finger between the strips, you'll have no trouble convincing yourself that there are two of them. This being the case, you might expect that cutting them both down the middle would produce two double length bands with four twists in each.

This guess turns out to be completely incorrect. When you split a double Moebius band down its center you get a pair of bands that are linked in a rather complicated way. The explanation for this can be found by examining the double Moebius band more closely. Take it apart, and you'll find that it isn't two nested strips at all but one long single strip with four twists in it.

All of this will raise a number of questions in the mind of the inquisitive child. Suppose a triple Moebius strip were cut down the middle. What would happen then? And what should we expect if we split a double strip with two twists in it? Finding answers to questions like these can provide hours of interesting experimentation.

A close relative of the Moebius strip is the *Klein bottle*. A Klein bottle has no edges and only one side. Its inside and its outside are one and the same. It is usually pictured this way.

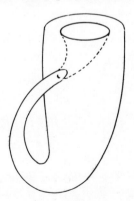

Theoretically, a Klein bottle can be made from a paper rectangle just as a Moebius strip can. One pair of opposite sides of the rectangle must be put together with a half twist, the other pair with no twist at all. This may sound simple enough. But the Moebius strip is a rectangle with one pair of opposite sides joined in a half twist. If you try to join the other two sides—that is, connect the edge of the Moebius strip to itself—you'll find that it's absolutely impossible.

One solution to this problem was discovered by Stephen Barr, the author of a book on topology. His idea is to start with a cylinder. To make one, cut out a rectangle and tape its long sides together.

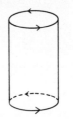

Now a half twist must be made before the other pair of edges are joined, and that means that when the arrows in the picture are brought together, they must be going in opposite directions. This is harder to manage than it might seem. Twisting the cylinder itself won't affect the relative directions of the arrows. It's easy to see this for yourself if you draw some arrows going in the same direction around the top and bottom of your cylinder.

Instead, you must make a slit at one end of the cylinder and pass the other end through it.

Now the arrows show that the edges have been twisted. Tape them together as best you can to form a single lip.

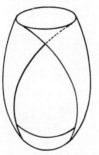

The slit in the Klein bottle really shouldn't be there. (In a four-dimensional space it wouldn't be necessary.) You must try to imagine that the bottle's surface is not interrupted at this joint.

If you fold the Klein bottle flat, as shown in the picture below,

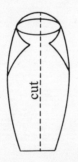

and cut it along the dotted line, you'll get two Moebius strips with their half twists going in opposite directions. Of course the reverse should be possible, too; you should be able to put it back together. As some anonymous author has written:

A mathematician named Klein
Thought the Moebius strip was divine.
Said he, "If you glue
The edges of two
You'll get a weird bottle like mine."

Can you do it?

VI

Flexagons

A *flexagon* is a bit of now-you-see-it-now-you-don't magic that's done with a strip of twisted paper. In one of its simplest forms, it has three faces that show themselves two at a time.

Transfer this pattern to a piece of heavy paper, cut it out carefully, and fold both ways along each of the lines.

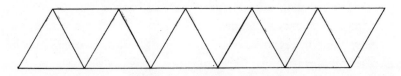

Now have your child color both sides of the strip according to the scheme shown below. The blank on the left should have green on the back of it.

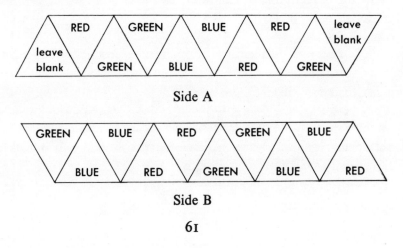

Side A

Side B

To put the flexagon together, fold the left blank over onto the red triangle next to it. On the other side of the strip there are two adjacent red triangles. Fold one of these over the other, and do the same with the two red triangles on the first side. Now the flexagon looks like this:

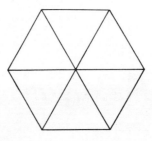

One side is all green; the other has five blue triangles and a red one. Tuck this red triangle behind the blue one underneath it, letting the green triangle on the other side hang over it as a sort of flap. The back of this flap is blank, and so is the back of the red triangle. Glue the two blanks together.

Now your flexagon is ready for flexing. Press two neighboring triangles downward until they are touching back to back, and push the opposite corner down to meet them. This will automatically bring a new face into view. (If the flexagon won't flex with the pair of triangles you have chosen, move on to the next pair.) When you start with green on the top and blue on the bottom, the flexing will produce a red top and a green bottom. The blue face will disappear completely!

This flexagon is called a *hexaflexagon* because it has six sides. When mathematicians first began to study hexaflexagons, they were quick to realize that this one was only the beginning. They called it a trihexaflexagon because it has three faces. A tetrahexaflexagon, by contrast, has four. The pattern for making it, with the colors for both sides, is shown below. As before, transfer it, cut it out, fold along each of the lines, and color.

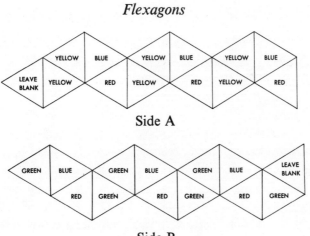

Side A

Side B

Now fold each yellow triangle over the yellow triangle next to it. This will give you a row of ten triangles, just like the pattern for the trihexaflexagon. Fold it up the way you did before, and glue the blanks together.

Here is the pattern for the pentahexaflexagon, with five faces.

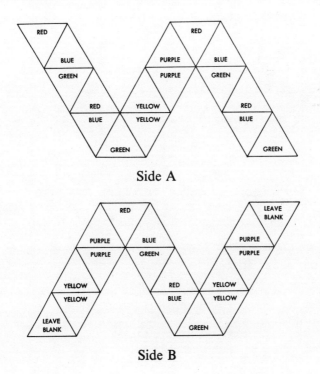

Side A

Side B

Fold all the yellow triangles onto the yellow ones next to them, and all the purple ones onto the purple. This will give you the familiar row of ten triangles to fold up as before.

Even though the pentahexaflexagon has five different faces, you may get into a sort of flexing rut in which the same three keep coming up again and again. The best way to get out of it is to keep flexing with the same pair of triangles as long as you can and then move on to the next. In this way, you will bring out all the faces eventually, although some will appear more often than others.

You can use the trihexaflexagon pattern to make a hexaflexagon, but you'll have to use it twice in order to make a row of nineteen triangles. The arrangement is shown below, with colors indicated for both sides.

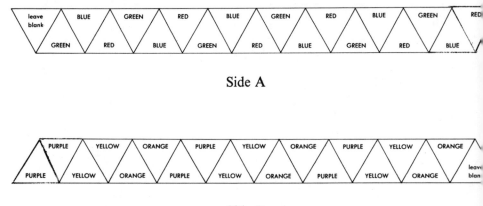

Side A

Side B

Fold orange on orange, yellow on yellow, and purple on purple. This will serve to wind the strip up until it looks like the original pattern for the trihexaflexagon. Fold it as before.

There are two other ways to make a hexahexaflexagon. In fact, from this point on, the more faces a hexaflexagon has, the more ways there are to design it.

Next for those with hexaflexafever, is a pattern for a heptahexaflexagon.

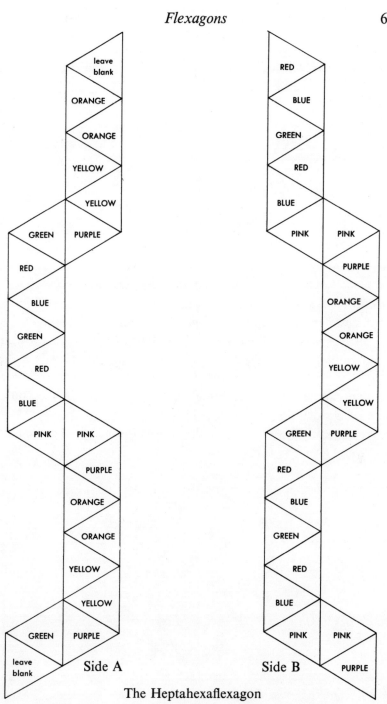

Side A

Side B

The Heptahexaflexagon

If you fold all the pink triangles onto the pink ones next to them, the heptahexa will look like the hexahexa, and you can fold it up exactly as you did before.

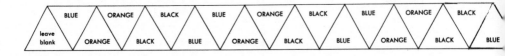

Wind it up by folding orange on orange, black on black, and blue on blue. Fold up the ten remaining triangles in the usual way, tuck in the flap on the end, and glue.

A hexaflexagon with twelve faces can be made in much the same way as one with six. Start with a strip just twice as long, and wind it up until it has the same shape as the hexahexaflexagon did originally. Then proceed as you did before.

You can at least double the number of different faces on a hexaflexagon by drawing designs instead of using solid colors. This is because each flexing changes the relative positions of the sides.

You might try a star,

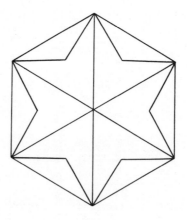

or a flower,

or something of your own invention.

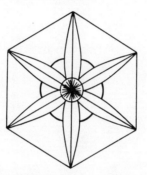

Here is the way they will look when you flex them:

If you use different colors, the effect is even more dramatic.

The best way to make designs like these is to leave the hexaflexagon blank until you have folded it up and glued it together. Then draw your designs, and watch them spin around.

Hexaflexagons have six sides; their cousins, the *tetraflexagons*, have four. To make a tritetraflexagon (three faces, four sides), use the pattern and the color arrangement shown below. (If you use graph paper, you won't have to trace the pattern.)

BLUE	BLUE	GREEN		LEAVE BLANK
		GREEN	RED	RED

Side A

		BLUE	BLUE	GREEN
LEAVE BLANK	RED	RED		GREEN

Side B

Hold Side A toward you, and fold its two blue squares together so that the flexagon looks like this:

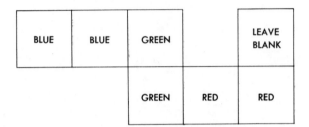

BLANK	GREEN		BLANK
	GREEN	RED	RED

Fold the red square behind the green one next to it so the flexagon looks this way:

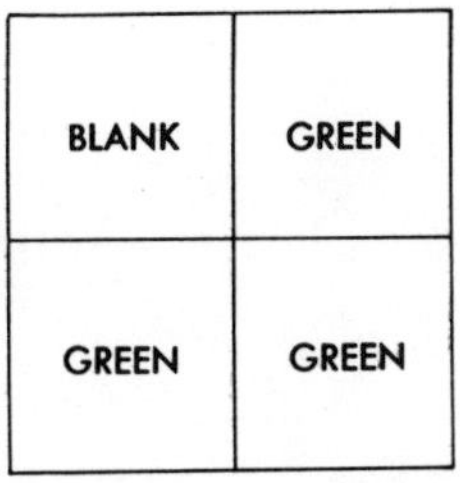

BLANK	GREEN
GREEN	GREEN

The blank square is now on top of a green one. Tuck it in back, glue the two blanks together, and you're ready to flex.

The only side of the tritetraflexagon that flexes is the green one. Fold it backward along the vertical line down its center, and the blue face will appear. Repeat the procedure, always on the green side, and you'll see the red face.

To make a tetratetraflexagon, cut out a block of twelve squares, three by four, color it according to the pictures below, and cut along the dotted lines.

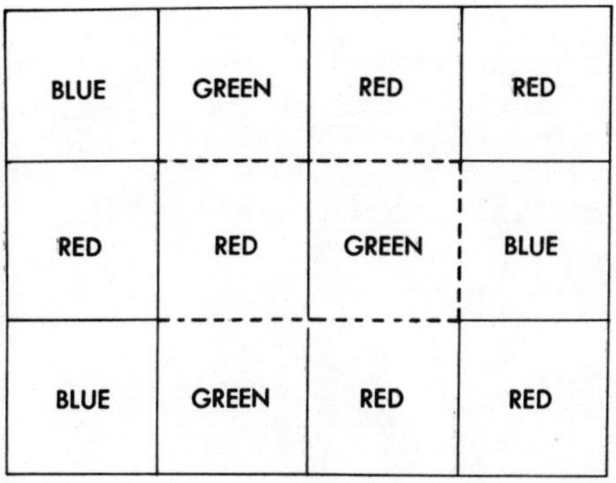

BLUE	GREEN	RED	RED
RED	RED	GREEN	BLUE
BLUE	GREEN	RED	RED

Side A

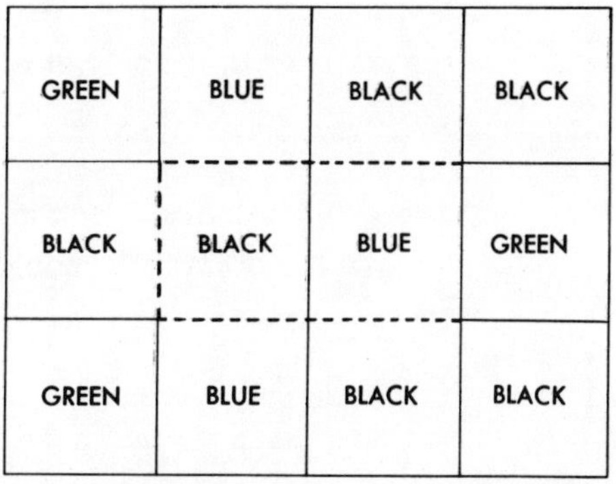

GREEN	BLUE	BLACK	BLACK
BLACK	BLACK	BLUE	GREEN
GREEN	BLUE	BLACK	BLACK

Side B

Holding Side A toward you, fold all the red squares onto the red ones next to them. Now the flexagon should look this way:

	BLUE	GREEN	GREEN
BLACK	BLUE		BLACK
	BLUE	GREEN	GREEN

Stick a piece of transparent tape to the black square on the left, as shown below,

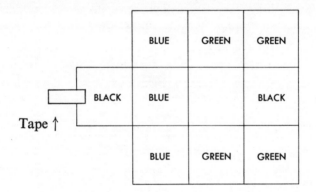

Tape ↑

fold the green squares over the green, and turn the flexagon over. This is how the other side should look.

Fold green on green again, and use the tape to attach the two middle squares together. The side of the flexagon facing you is now all black; the other is all blue.

The tetratetraflexagon flexes the same way the tritetraflexagon does. The blue face bends back along its center to reveal first the green and then the black, and the green one bends back to show the red and the blue.

This is a pattern for a flexagon with four sides and six faces, a hexatetraflexagon. Cut it on the dotted line.

BLUE	RED	PURPLE	PURPLE
BLUE			GREEN
GREEN			BLUE
PURPLE	PURPLE	RED	BLUE

RED	ORANGE	BLACK	GREEN
BLACK			ORANGE
ORANGE			BLACK
GREEN	BLACK	ORANGE	RED

Side A   Side B

Fold the purple squares on the purple squares next to them, and the blue on the blue. Now one side of the flexagon should look like this:

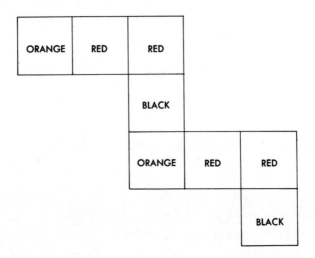

Fold the red squares on the red ones next to them to get this, and stick a piece of transparent tape to the under side of the green square at the bottom.

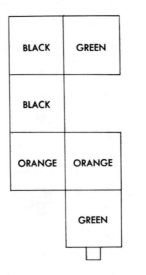

← Tape

Then turn the flexagon over and fold green on green.

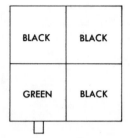

Tuck the green square behind the black one underneath it, fold the tape over the edge, and attach it to the other side. This will seal the flexagon.

The black side that faces you now won't flex, but the orange one on the other side will, both horizontally and vertically. If you experiment a little, you should have no trouble bringing out all six of the faces.

Mathematicians know how to make many other tetraflexagons, but they don't understand them nearly as well as they do the hexaflexagons. Tetraflexagons come in a bewildering variety of shapes and sizes, and so far no one has developed a theory that explains all of them and shows how they are related to one another. Maybe one of today's young people, whose interest is captured by the problem, will be the first to solve it.

VII

A Visit to Flatland*

THE FOURTH DIMENSION! How romantic and mysterious those words sound. Is there a path, unknown to us, that leads out of space as we know it? If there is, it must go through space of an entirely different kind.

Is there any reason why the number of dimensions should be limited to three? Suppose a four-dimensional world really existed. How would it differ from ours? Questions like these are as fascinating as they are impossible to answer. Our experience has been limited to three dimensions, and our imaginations boggle at the thought of more.

There is only one way to study four-dimensional space, and that is by analogy. Since we are familiar with three dimensions, it is easy for us to imagine a space with only two. And by studying such a space in relation to our own, we can deduce a surprising amount about higher dimensional spaces.

In 1884 the Reverend Edwin Abbott, a British schoolmaster whose hobby was mathematics, wrote a little book called *Flatland*, which has become a modern classic. A forerunner of science fiction, it describes life in a perfectly flat world peopled by lines, polygons, and circles. The version that follows was adapted especially for young readers.

* Adapted from *Flatland*, by Edwin A. Abbott.

74

PART I

This World

I call our world Flatland, not because we use that name for it, but to make its nature clearer to you, my happy readers who are privileged to live in space.

Imagine a vast sheet of paper covered with straight lines, triangles, squares, pentagons, and other figures. The figures are free to move about on the paper, but they can never rise above it or sink below it. This will give you a good idea of my country and my countrymen. A few years ago I would have said "my universe," but now my mind has been opened to higher things.

In such a country you can see that it is impossible to have anything that you would call a solid, but you might think that we could tell the difference between triangles, squares, and other figures. On the contrary, we can do nothing of the sort.

Cut a circle, a square, and a triangle out of paper, and put them on a table. As long as you look down on these shapes, you get a three-dimensional view of them. To see them as a Flatlander would, you must lower your eyes to the table's edge.

Watch the shapes carefully while you do this, and you will see all of them change. The circle becomes an oval. The square turns into a rectangle. The triangle grows thinner and thinner. By the time your eye reaches the level of the table, all three have turned into lines. In two dimensions it is impossible to tell one from another.

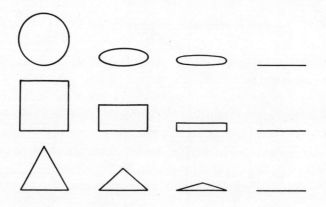

Straight lines are all we can ever see in Flatland. When our friends come close to us their lines grow larger, when they leave us their lines grow smaller again, but they all look alike whether they are circles, squares, or triangles.

You might wonder how, under such difficult circumstances, we are able to tell one of them from another, but I can explain that better when I describe the inhabitants of Flatland. Right now let me tell you something about our climate and our houses.

Like you, we have four points of the compass: north, south, east, and west. Without any sun or stars to guide us, we cannot determine north the way you do, but we have a method of our own. In Flatland there is a constant attraction to the south, and although in some parts of the country it is very slight, its hampering effect is usually enough to serve as a compass. We are also aided by the rain, which always falls from the north.

Our houses are built with their side walls running north and south so that the roofs can keep the rain out. There is no need for windows, because it is always light in Flatland, both indoors and out. No one knows where the light comes from, although many of our learned men have tried to find out. Alas, I alone in all of Flatland know the answer to this mysterious problem, but I cannot make a single one of my countrymen understand it. I am mocked and ridiculed, I, the only one who knows the truth about space and the introduction of light from the world of three dimensions. But enough of this painful subject; let me return to our dwellings.

Most of our houses are five-sided, or pentagonal. The two northern sides are the roof, and the southern side is the floor. On the eastern side there is a small door for women; on the western, a large one for men.

Square and triangular houses are outlawed in Flatland because their angles are so sharp. Houses have much dimmer lines than people do, and there is always the danger that an absent-minded traveler might injure himself by running into the point of an angle.

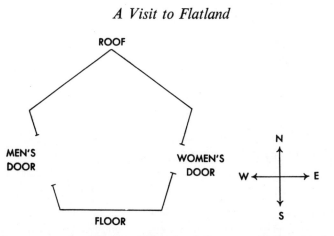

ROOF

MEN'S
DOOR

WOMEN'S
DOOR

FLOOR

N

W ← → E

S

The greatest length or width of a full-grown Flatlander is about twelve inches. Our women are straight lines. Our soldiers and workmen are triangles with two equal sides, each about eleven inches long, and a third side no bigger than half an inch. This small third side makes the angle opposite it sharp and formidable. We call these triangles isosceles, just as you do.

Our middle class consists of equilateral triangles. Professional men are squares, to which class I myself belong, and pentagons. Above these come the noblemen, beginning with the hexagons and increasing in the number of their sides until they receive the honorary title of polygonal. When the number of sides becomes so great, and the sides themselves so small, that the figure cannot be distinguished from a circle, he is included in the circular order, the highest class of all.

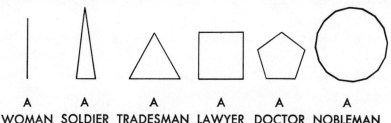

A   A   A   A   A   A
WOMAN SOLDIER TRADESMAN LAWYER DOCTOR NOBLEMAN

If the highly pointed angles of our soldier class are formidable, our women are far more so. For if a soldier is a wedge, a women is a needle. Add to this her power of becoming invisible whenever she pleases, and you will see that in Flatland a woman is by no means to be trifled with.

Perhaps you are wondering just how our women can make themselves invisible. Put a needle on a table and lower your eyes to the table's edge. When you look at the needle sideways, you can see its whole length, but try turning it around slowly. It will appear to shrink until finally it is nothing but a tiny, gleaming point. And that is exactly the way it is with our women. When their sides are turned to us we see them as lines, but when they turn around they disappear completely.

You can see at once how dangerous such a situation can be. When a woman is invisible, it is often impossible to avoid colliding with her. And to run into a woman means absolute and immediate destruction.

Many laws have been passed to safeguard our men from this hazard. Every house has a special entrance for women on its eastern side; under no circumstances may a woman use the men's door on the west. In public places a woman must keep up a peace cry, while swaying constantly from left to right so that she can always be seen. And any woman certified to be suffering from St. Vitus' dance, fits, or a chronic cold accompanied by violent sneezing is destroyed at once.

Above all, a woman is not to be irritated as long as she is in a position where she can turn around. Only when she is in her home, which is designed with a view to denying her that power, can you do and say as you like.

Now I come to our methods for recognizing one another. In the world of three dimensions, you are blessed with shade as well as light, and you also have a knowledge of perspective. You can actually *see* an angle and look at the complete circumference of a circle. How can I make it clear to you the extreme difficulty which we in Flatland have in telling one another apart?

We know our friends by their voices, of course, but this method of recognition has its disadvantages. An isosceles can easily feign the voice of a polygon and, with some practice, of a circle himself. To avoid confusion, we use the method of feeling.

Feeling is, among our women and lower classes, the principal test of recognition. What an introduction is for you, the process of feeling is for us. "Let me ask you to feel Mr. So-and-so," is what we usually say when presenting one Flatlander to another.

The process of feeling is not nearly as difficult as you might think. We do not have to feel all the sides of an individual to find out how many he has. As a rule, we can determine this by feeling just one of his angles. Long practice and training, begun in the schools and continued in the experience of daily life, enables us to distinguish at once between the angles of an equilateral triangle, a square, and a pentagon.

With the nobility, though, it is much harder. Even a Master of Arts in our university has been known to confuse a ten-sided polygon with a twelve-sided one, and there is hardly a Doctor of Science in or out of that famous institution who could pretend to decide promptly between a twenty-sided and a twenty-four-sided member of the aristocracy.

Feeling also has its dangers. To avoid the possibility of being injured by a sharp angle, it is essential for the feeler that the felt should keep perfectly still. A start, a fidgety shifting of position, or even a violent sneeze has been known to prove fatal and nip many a promising friendship in the bud.

Another method of recognition is used by the higher classes in the temperate zones of our country. In these zones there is always a heavy fog. Without the fog, all lines would look equally clear. But when the fog is heavy enough, objects that are, say, at a distance of three feet look dimmer than those at a distance of two feet eleven inches, and this difference makes possible the art of sight recognition.

Suppose, for example, that I see two individuals coming toward me. Let us say they are a merchant and a doctor or, in other words, an equilateral triangle and a pentagon. I have no way of knowing this, of course. How can I tell them apart?

All I have to do is look directly between two of their sides. In the case of the merchant, I see a straight line, very bright at the center but dimming off rapidly to either side. The doctor looks like a straight line too, but his sides do not dim off nearly as rapidly because his angles are so much larger. These differences in dimness tell me which is which.

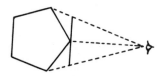

The Doctor   The Merchant

After long training and constant experience, a well-educated Flatlander is usually quite good at sight recognition, but it is never an easy matter. If a triangle should happen to present his side to me, for instance, I would either have to ask him to rotate, or else edge my eye around to one of his angles. Until I had done one or the other, I would not be able to tell him from a woman.

Everything I have said about feeling and sight recognition depends upon one single fact: that all the inhabitants of Flatland are regular figures. Women are not just lines, they are straight lines. Soldiers always have two of their sides equal. Tradesmen have three equal sides, lawyers and doctors four, and it is the same with the other polygons. All of their sides must be equal.

The length of the sides depends upon age. A newborn baby girl is about one inch long; a full grown woman might be almost a foot. And as far as the males are concerned, all of their sides added together are usually three feet or a little bit more. But I am not speaking of the size of our sides here; I am speaking of their equality. Our entire way of life depends on it.

If our sides were unequal, our angles would be unequal. We would not know a person's shape from just one of his angles; we would have to feel each and every one of them. Sight recognition would be impossible, feeling would become an endless process, and it would be dangerous for us to move about at all.

Suppose I were to meet a tradesman in the street and invite him home for lunch. As long as his three sides are equal, I can do this with perfect confidence, because everyone knows the area of an adult triangle. But suppose he were dragging a parallelogram around behind his vertex. What would I do with a monster like that stuck in my doorway? Irregularity such as this is incompatible with the safety of our state.

Flatland is governed by the Council of the Circles, which makes all the important decisions about business, art, science, trade, architecture, and education.

Of course no circle is really a circle; he is just a polygon with a very large number of very small sides. As the number of sides increases, the polygon looks more and more circular, and when this number is very great indeed, say three or four hundred, it is extremely difficult for even the most delicate touch to feel any angles. I should really say it *would* be difficult, for to feel a circle would be an insult of the worst sort. This makes it easy for a circle to keep the exact nature of his perimeter or circumference a secret.

But enough of this. Now that I have told you something about Flatland, it is high time that I go on to the central event of this book, my initiation into the mysteries of space.

PART II
Other Worlds

It was the next to the last day of the 1999th year of our era. I had amused myself until late at night with geometry, my favorite pastime, and had gone to bed with an unsolved problem on my mind. During the night I had a strange dream.

I saw before me a tremendous number of straight lines, which I naturally thought must be women. Between them were other creatures who looked like tiny, lustrous points, and all of them were moving back and forth in the same straight line.

I approached one of the largest lines and tried to speak to her, but she didn't seem to hear me. I tried again, and then again, but all to no avail. Losing patience with what seemed to be thoroughly bad manners, I brought my mouth directly in front of hers and repeated my question loudly.

"Woman, what is the meaning of this monotonous motion back and forth in a straight line?"

"I am not a woman," replied the line, "I am the king of the world. Where did you come from, and what are you doing in Lineland?"

I begged pardon if I had startled His Royal Highness and, describing myself as a stranger, asked him to tell me something about his country. But I had the greatest difficulty in getting the information I wanted, because the king seemed to think that whatever was familiar to him was known to me as well. However, I did succeed in discovering the following facts.

It seemed that this poor king, as he called himself, believed that the straight line which he called his kingdom was the whole of space. Not being able to move or see except in his line, he couldn't imagine anything outside it. He had heard my voice when I first spoke to him, but he didn't see me, and the sound had seemed to come from inside him. He couldn't imagine where I had come from; outside of his line nothing existed for him.

His subjects were also confined to the straight line that was their world. No one could ever move over, and no Linelander could get past the two on either side of him. Once neighbors, always neighbors.

"Don't you find it terribly dull moving back and forth in the same straight line all your life?" I asked the king. "Why, all you can ever see is a point."

"Of course," said the king, "but my subjects are both points and lines."

"How can Your Highness tell the difference?" I asked him. "I myself *saw* many of your points and lines before I entered Lineland."

"Don't be ridiculous," replied the king. "Everyone knows it is impossible to see the difference between a point and a line. One must hear it. Take me for example. I am the longest line in Lineland. I take up over six inches of space."

"Of length," I corrected him.

"Fool!" he cried, "Space is length. One more interruption and our conversation is at an end."

"Forgive me," I said, "but I would be most interested to know how length can be heard, as you put it."

"Every line has two voices," explained the king, "one at each end. When I call out with both of my voices at once, the second follows the first in exactly the time it takes for sound to travel the six inches of my length. This calculation can be made by anyone who is interested."

"But if you would only move out of your line to the left or the right, you would be able to see one another."

"I have no idea what you mean by left and right. I suppose those are the words you use for north and south."

"Oh no," I said. "Where I come from, besides moving northward and southward we can move to the left and the right."

"Show me, please, this motion to the left and the right."

"That I cannot do, sir, unless you step out of your line altogether."

"Out of my line? Do you mean out of the world? Out of space?"

"Well, yes," I said, "out of *your* world and *your* space. But your space is not the true space. True space is a plane, and your space is only a line."

"I have no idea what you mean by a plane."

"It's a very difficult thing to explain," I told him. "Let me see now. When you are moving back and forth in your line, does it ever occur to you that you might move in some other way? Don't you ever want to move in the direction your side faces?"

"Never. How can my inside 'face' in any direction? And how could I possibly move in the direction of my inside?"

"What you call your inside is just the part of you that you cannot see. But words will get us nowhere. Let me move out of Lineland in the direction I am trying to describe to you."

I began moving slowly out of Lineland, but as long as any part of me remained in his line the king kept exclaiming, "I see you, I see you. You are nothing but a line, and you are not moving a bit."

"I am a square," I cried, but the king had no idea what I meant by that.

Then, when I had finally moved out of his line completely, the king cried, "You are gone. You have disappeared!"

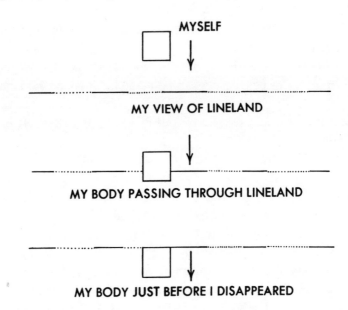

"I have done nothing of the sort," I replied. "I am simply out of Lineland, that is to say, out of the straight line you call space, and in the true space where I can see things as they really are. At this very moment I can see all of your line, or what you would call your inside, and all of the points and lines on either side of you. Are you convinced now?"

"If you had any sense," replied the king, "you would listen to reason. You ask me to believe that there is another space besides mine and that you are able to move about in it. But when I ask you to show me this motion, you do nothing but disappear into thin air."

"It is you who will not listen to reason," I cried. "You say you can see when you have nothing to look at but a point. Lineland can hold only points and lines, but where I come from there are triangles, squares, pentagons, hexagons, and even circles. You are a line, but I am a line of lines. Really, you are quite hopelessly dull."

When the king heard that, he came rushing toward me, and a war cry arose from his subjects. Paralyzed with fear, I listened as the cry grew louder and louder and awoke to find the breakfast bell welcoming me back to Flatland.

From dreams I go on to facts.

It was the evening of the last day of the 1999th year of our era. My sons and grandchildren had gone to their rooms, and I was sitting with my wife, remembering the year just past and looking forward to the one ahead.

I was also thinking about something my youngest grandson had said. He was an unusually bright young hexagon, and his uncles and I had been giving him his daily lesson in sight recognition. He had done so well that I had decided to reward him with some lessons on the applications of arithmetic to geometry.

I took nine one inch squares and put them together to make a large square with a side of three inches. Then I explained to my grandson that although it was impossible for us to *see* the inside of a square, we could find the number of square inches in it by squaring the number of inches on a side.

"Since 3^2 is 9," I said, "9 is the number of square inches in a square whose side is 3 inches long."

The little hexagon thought for a minute and then he said, "But you have also been teaching me to raise numbers to the third power. 3^3 is 27. That must mean something in geometry, too."

"Nonsense," I said. "How can it mean anything when there are only two dimensions?"

Then I showed him how a point, moving in a straight path for three inches, makes a line three inches long which is represented by 3, and how this line, moving parallel to itself in another straight path for three inches, makes a square three inches on a side which is represented by 3^2.

At this my grandson suddenly exclaimed, "Well then, if a point moving three inches makes a line we call 3, and if this line moving three inches parallel to itself makes a square we call 3^2, then this square moving three inches parallel to itself, though I can't imagine how, must make something else, though I don't know what, that we would call 3^3."

"Stop being foolish and go to bed," I said, and my little grandson disappeared.

Only a few sands remained in the hour glass now, and I turned it northward for what would be the last time in the old millennium. As I did, I exclaimed out loud, "The boy is a fool!"

Immediately I sensed a strange presence in the room, and a chill swept over me.

"He is no such thing," cried my wife, but I paid no attention to her. I searched the room and found nothing, but the chill swept over me again.

"What are you looking for?" asked my wife. "There's no one here."

She was right, and I sat down saying, "The boy is a fool! 3^3 can have no meaning in geometry."

A reply came at once. "The boy is not a fool, and 3^3 has an obvious geometrical meaning."

My wife and I both heard these words, and springing forward in the direction of the sound, we were horrified to see a figure! At first glance it seemed to be a woman, but I could see that its ends dimmed too rapidly for it to be a straight line. My wife, however, did not notice this.

"How did she get in here?" she cried. "You promised me that there would be no ventilators in our new house."

"And I kept my word," I said. "Tell me, what makes you so sure that the stranger is a woman? I can see by my power of sight recognition . . ."

"I have no faith in your sight recognition," said my wife. "Feeling is believing."

"Well," I said, "if you are so certain, why don't you demand an introduction?"

Assuming her most gracious manner, my wife approached the stranger.

"Good evening madam," she said. "Permit me to feel and be felt by . . . ," but she stopped in the middle of her sentence. "It is not a woman," she cried, "and there are no angles either, not a trace of one. Can it be that I have misbehaved in front of a perfect circle?"

"In a certain sense I am a circle," came the reply, "and a more perfect circle than any in Flatland, but to put it more accurately, I am really many circles in one."

Then the stranger addressed himself to my wife. "I have a message for your husband, dear lady, that I cannot deliver in your presence. If you would excuse us for a few moments . . ."

My wife hastened to assure our distinguished guest that her bedtime had long since passed, and with many apologies for her indiscretion, she went to her room. I glanced at the hour glass again. The last sands had fallen. The new millennium had begun.

When my wife had left us, I turned my full attention to the stranger. His appearance was astonishing. He seemed to change his size from one moment to the next in a manner quite impossible for any figure I had ever seen before. The thought flashed through my mind that he might be a monstrous isosceles, feigning the voice of a circle and preparing to stab me with his acute angle. Desperate with fear, I rushed across the room to feel him. My wife had been right. There was not the slightest trace of an angle. Never had there been a more perfect circle. The stranger stood very still while I walked around him, beginning at his eye and returning to it again.

"Have you felt me enough now?" he asked me. "Are you quite introduced?"

"Forgive me," I begged him, filled with shame at the thought that I had been impertinent enough to feel a circle, "and please tell no one about this, especially my wife. I am honored to have you here, sir. It is just that your unexpected visit has surprised me and made me a little nervous. May I ask you, kind sir, where you have come from?"

"From space, of course. Where else?"

"Pardon me, sir, but is it not true that both of us are in space at this very moment?"

The stranger's voice was full of scorn. "What do you know about space? Define it."

"Space sir, is length and width."

"You see. You do not even know what space is. You think of it as having only two dimensions, but I have come to tell you of a third. There are not only length and width. There is height."

I laughed, thinking that he was making some sort of joke. "We talk of height and breadth, too, sir, and of length and width, but all four words are names for the same two dimensions."

"But I am using three names for three *different* dimensions."

"Will you explain to me in what direction I can find this third dimension?"

"It is the direction from which I came. It is above you and below you."

"Do you mean that it is northward and southward?"

"I mean nothing of the sort. I mean a direction in which you cannot see because you have no eye in your side."

"Forgive me, sir, but if you will look at me carefully you will see that I have a perfect eye at a point where two of my four sides meet."

"Yes, but in order to see into space, you would need an eye, not on your perimeter, but on your side, that is, on what you would probably call your inside but what we in Spaceland would call your side."

"An eye on my inside! In my stomach! You must be joking."

"I am in no joking mood. I tell you that I come from space or, since you do not seem to know what I mean by that, from the world of three dimensions, where I can see things as they really are. You cannot see inside things that are enclosed on all sides, but just now, from a position above the plane you call your space, I looked down and saw your buildings and your people, and I saw their insides and their outsides all at the same time."

"That is a claim anyone can easily make, sir."

"But not easily prove, you mean. Well, I intend to prove it. When I looked down on your house before, I saw your four sons, the pentagons, and your two grandsons, the hexagons. I saw your youngest grandson stay with you for a while and then go to his room, leaving you and your wife alone. Then I came here, and how do you think I got in?"

"Through the roof, I suppose."

"You know as well as I do that your roof was just repaired so that not even a woman could get in. I tell you I came from space. Are you not convinced by what I have told you about your family?"

"Such information can be obtained in many ways."

"Listen to me!" cried the stranger. "You are living on a plane. What you call Flatland is a vast, level surface on which you and your countrymen move about without ever rising above it or falling below it. I am not a plane figure but a solid. You call me a circle, but I am really not a circle. I am an infinite number of different sized circles, one on top of another. The smallest is just a point, and the largest is thirteen inches in diameter. Where I come from, I am called a sphere. There is not enough room in your two-dimensional world to hold all of me at once, so as I pass through it, as I am doing right now, you see only a slice, or section, of me, which you call a circle."

"That reminds me of a dream I had last night," I told him. "It was about a country called Lineland, where they had only one dimension. Just one of my lines could pass through it at a time. The king himself didn't believe I was a square. In fact, he had no idea what a square was."

"Just as you have no idea what a sphere is," said the stranger. "You can see only one of my circles at a time because you can't raise your eye out of the plane of Flatland. But at least you can see my sections growing smaller as I rise in space. Watch now."

I could see no "rising," but the circle began to shrink, just as he had said he would, until finally, he vanished completely. I blinked once or twice to make sure I wasn't dreaming. Then suddenly, from the depths of nowhere, came a hollow voice asking, "Am I quite gone? Do you believe me yet? Now I will lower myself into Flatland again, and you will see me grow larger and larger until I am right back where I started from."

Every reader in Spaceland will easily understand what my visitor was doing, but for me it was by no means so simple. A rough diagram will make it clear to any Spaceland child that a sphere, rising through Flatland, must appear to the Flatlanders as a circle, first large, then smaller, and then very small indeed. But to me, although I saw the facts before my eyes, the explanation was as mysterious as ever. All I could see was that the circle had made himself smaller and smaller until he had disappeared, and then had reappeared and made himself larger and larger again.

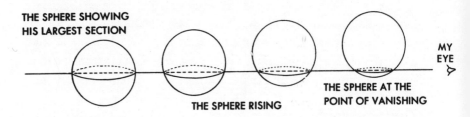

THE SPHERE SHOWING HIS LARGEST SECTION

THE SPHERE RISING

THE SPHERE AT THE POINT OF VANISHING

MY EYE

When he was back to his original size, he heaved a great sigh, because he realized that once again I had completely failed to understand him. And I, for my part, was beginning to believe that the old wives' tales were true and that there might be such things as witches and sorcerers after all.

Finally I heard him mutter to himself, "There's only one thing left. I must try the method of analogy."

Addressing himself to me again, he said, "Tell me, if a point moves directly northward, what name would you give to its path?"

"A straight line."

"And how many ends does a straight line have?"

"Two."

"Now imagine the line moving eastward through a distance equal to its own length. What would you call the path now?"

"A square."

"And how many sides does a square have? And how many corners?"

"Four sides and four corners."

"Now stretch your imagination a little and think of the square moving parallel to itself in an upward direction."

"Do you mean northward?"

"No, not northward, upward. Out of Flatland altogether. If it moved northward its southern points would pass through places its northern ones had already occupied. That is not what I mean at all. I mean that every point in you—for you are a square and will serve my purpose perfectly—that every point in you, that is to say in what you would call your inside, is to move in such a way that no point passes through a place previously occupied by another. Each point moves in a straight line all its own."

Restraining my impatience with what seemed to me a perfectly outrageous suggestion, I asked, "And what sort of figure results from this peculiar motion?"

"It is not a figure at all, but a solid, and you should be able to answer that question for yourself. *One* point produces a line with *two* end points. The line produces a square with *four* end points. 1, 2, 4. Each number is just twice as great as the one before it. What number comes next?"

"Eight."

"Exactly. The square produces something called a cube, which has eight end points, or vertices."

"And has this creature sides as well as terminal points?"

"Of course, but not what *you* would call sides, what *we* would call sides. You would probably call them solids."

"And how many sides, or solids, will this creature have whom I am to produce by moving my inside in an 'upward' direction?"

"Need you ask? If a point has no sides, a line two, and a square four, what must we say of a cube? 0, 2, 4. Each number is two more than the one before it. What number comes next?"

"Six."

"Exactly. So you see, you have answered your own question. The cube will be bounded by six sides, that is, by six of your insides. You see it all now, eh?"

Suddenly I could stand no more. I rammed into the stranger with my hardest right angle, using a force that would have destroyed any ordinary circle, but I could feel him slipping slowly away from me. He did not move forward or off to one side, but somehow out of the world completely, vanishing into nothing. Then the hollow voice came again.

"Why do you refuse to listen to reason? Only once in a thousand years am I allowed to reveal the third dimension to a Flatlander. I chose you because you seemed to be a reasonable man, but it seems impossible to convince you. Wait! Maybe actions will speak louder than words.

"Listen, my friend. I have told you that from my position here in space I can see the inside of everything in Flatland. That closet, for instance. There are some money boxes in it, and two account books. Now I'm going to lower myself into the closet and bring you one of those books. I saw you lock it up half an hour ago, and I know you have the key. But that won't stop me; I can get in even though the door is locked. I'm inside the closet now, and I'm taking the book. Now I have it, and I'm carrying it off to Spaceland."

I rushed to the closet and threw the door open. One of the account books was missing! With a mocking laugh, the stranger appeared in another corner of the room, and at the same time the account book appeared on the floor.

"Surely you must believe me now," he said. "My explanation is the only one that fits the facts. What you call space is nothing but a plane. I am in the real space, and I can look down on the insides of things that are closed to you. You could leave your plane if you really wanted to, and then you would be able to see all the things I can. The higher I go, the more I see, although of course things look smaller and smaller.

"Right now I am looking at your neighbor the hexagon and his family. Now I see the inside of the theater on the next block. The audience is just leaving. On the other side of the theater, a circle is sitting in his study. And now I'll come back to you. As a final touch, what do you say I give you a gentle poke in the stomach? It won't hurt a bit, I promise you."

Before I could say a word, I felt a shooting pain, and a devilish laugh seemed to come from deep inside me. A moment later the stranger returned.

"There, that didn't hurt much, did it? If you aren't convinced now, you never will be."

I had reached the limit of my endurance. It was unbearable to be at the mercy of a magician who could play tricks with one's very stomach. If only I could manage to pin him against the wall until help came!

Once more I dashed my hardest angle against him, and at the same time I woke the whole household with my cries for help. I seemed somehow to have trapped the stranger outside of Flatland altogether, and he remained quite still while I pressed against him and continued my shouting.

A shudder seemed to run through the sphere. "This can't happen," I heard him say. "Either he listens to reason, or I use the last resort." Then, addressing himself to me again, he cried, "Listen! No one else must see what you have seen. Your wife is on her way in here. Send her back. Hurry, or I'll have to take you away with me to the world of three dimensions."

"Madman," I cried. "I'll never let you go."

"All right," he thundered, "you leave me no choice. Out of your plane you go. One! Two! Three!"

An unspeakable horror seized me, and everything went dark. Then I had a dizzy, sickening sensation of sight that was not like seeing. I saw lines that were not lines and space that was not space. I was myself, and I was not myself. When I could find my voice I shrieked, "This must be madness."

"It is nothing of the sort," replied the stranger calmly. "It is knowledge. You have entered the land of three dimensions. Try to look at it steadily."

I looked, and behold, a whole new world! There before my eyes was a perfect circle such as up to now I had only dreamed of. What seemed to be the stranger's center was open to my view, but instead of seeing his insides I saw a beautiful something for which I had no words, but which you, my readers in Spaceland, would call the surface of a sphere.

"How is it," I asked him, "that I can see your center but not your insides?"

"No one can see my insides," replied the sphere. "I am different from the inhabitants of Flatland. If I were just a circle, you would be able to see inside me, but I am made of many circles. And just as the outside of a cube is a square, my outside looks like a circle."

This explanation bewildered me completely.

"Don't worry if you have trouble understanding the mysteries of Spaceland at first," said the sphere. "Little by little they will become clear to you. Let's take another look at Flatland together. I want to show you something you have thought about a great deal but never really seen before: an angle."

"Impossible!" I cried, as the sphere led me along.

"There," he said. "Below you are your house and all the members of your family."

I looked down and saw things that until that moment I had only been able to imagine. There were my sons asleep in their rooms and my grandsons asleep in theirs. Only my wife, alarmed at my absence,

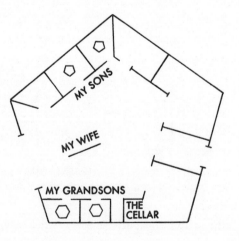

was up and about. As we drew nearer and nearer, I could even see what was inside my closets and cabinets. I was worried about my wife, and I wanted to go back and reassure her, but I found that I was unable to move.

"Don't trouble yourself about your wife," said the sphere. "She won't be alone for long. Come, I want to show you some more of Flatland."

Again I felt myself rising through space, and it was just as the sphere had said. The further away we went, the more I was able to see. There was my town, with the inside of every house and person completely open to my view.

"Look over there," said the sphere. "Do you recognize that building?"

I looked and saw an enormous polygon, which I realized was the General Assembly Hall of the States of Flatland.

"Here we descend," said the sphere.

It was morning by now, the first hour of the first day of the two thousandth year of our era, and the Council of the Circles was holding a solemn meeting, as it had on the first days of each of the eras preceding. The secretary was reading the minutes of the last meeting.

"Whereas the States of Flatland have been troubled on the first day of each era by ill-intentioned people pretending to have received revelations from another world and professing to give demonstrations which have provoked a state of frenzy in themselves and others, it is resolved that special injunctions be sent to the officers of each of the districts of Flatland to search for these people and commit them to the state asylum."

"You hear your fate," said the sphere. "You will be imprisoned if you try to reveal the third dimension to your countrymen."

"Oh no," I replied. "It is all so clear to me now that I think I could make a child understand it. Let me go down there right now and try."

"Not yet," said the sphere. "The time for that will come. Follow me. I have a mission to perform."

With these words, he lowered himself into the room where the circles were holding their meeting.

"I am here to tell you that there is a third dimension," he cried.

The younger circles were horrified to see the sphere's section widen before them. But on a signal from their president, who did not show the slightest surprise, six isosceles soldiers rushed upon the sphere.

"We have him," they cried. "Yes, we have him. No, now he's slipping away. He's gone!"

"My Lords," said the president to the other circles, "there is not the slightest need for you to be surprised. It says in the secret archives that this same thing happened on the first day of each of the last two eras. Of course you will say nothing about it outside this room. And now, our business being concluded, I have only to wish each and every one of you a happy new year."

I wanted to rush down and tell the circles all I had learned about the third dimension, but the sphere restrained me.

"Come along," he said. "I have more to show you before you return to Flatland."

Once more we ascended together.

"Until now," said the sphere, "I have shown you only plane figures. Now I must introduce you to solids and show you how they are made. Look at these square cards. I am going to pile them up, one on top of another, until the pile is just as high as it is wide. There, I am finished. That is what we call a cube."

I listened to the sphere as carefully as I could, but to me the "pile" of squares looked like this.

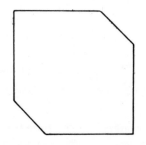

"Forgive me," I said, "but all I see is a figure very much like the ones that we have in Flatland."

"It only looks that way to you," said the sphere, "because you are not accustomed to light and shade and perspective, just as in Flatland a hexagon would look like a straight line to anyone who did not know the art of sight recognition. Feel the cube, and you will see that what I say about it is true."

I did as he asked, and discovered that the cube was not a plane figure at all, but a solid with six planes for its sides and eight corners called solid angles. When the sphere had taught me about "light," "shade," and "perspective," I was able to see the cube this way.

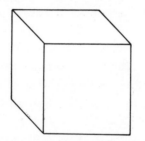

All this time I had been doing some serious thinking. Finally I said to the sphere, "I wonder, kind sir, if I might have a little peek at your insides."

"At my what?"

"At your insides. Your stomach and intestines."

"I told you before, no one can see inside me."

"But sir, if you yourself, who are superior to everyone in Flatland, combine many circles in one, there must be a being above you who is superior to everyone in Spaceland and combines many spheres in one. And just as you can look down on Flatland and see the inside of everything in it, there must be a place from which the insides of solid things can be seen."

"Stuff and nonsense. Come, I have much more to show you."

"Sir, please do not refuse me. One look at your insides is all that I ask."

"I would show them to you if I could, but believe me, it is quite impossible. I can't turn myself inside out."

"But you have shown me the insides of all my countrymen in the land of two dimensions by taking me with you to the land of three.

Why can't we go on another journey into the fourth dimension, so that we can see the insides of three dimensional beings?"

"And where do you suppose this land of four dimensions might be?"

"I don't know, but I'm sure you do."

"Ridiculous! You are quite mistaken. There is no such place."

"But there must be. If you had an eye in your stomach you would probably be able to see it. It is all according to the analogy that you yourself have taught me. If a line has two terminal points, a square four, and a cube eight, then the cube, when it moves in the direction of the fourth dimension, must produce something with sixteen terminal points. And if the line is bounded by two points, the square by four lines, and the cube by six squares, this 'something' must be bounded by eight cubes. Have none of your countrymen had visits from beings who entered their closed rooms the way you entered mine?"

The sphere paused for a moment. "There have been reports that they have," he said slowly, "but no one knows whether there is any truth to them."

"You see! I was certain of it. Tell me, did these beings also change their size the way you did and appear and disappear at will?"

"They disappeared, certainly, if ever they really appeared in the first place. But most people say that they weren't real, and that those who thought they saw them were only imagining things."

"I don't believe that," I told him. "But even if it is so, and this other space is only Thoughtland, let's go there together and see a cube moving in some altogether new direction to create something with sixteen terminal points and eight cubic faces. And from there we can travel on through a fifth dimension, and a sixth, and a seventh . . ."

How long I would have continued I don't know. The sphere, in a voice of thunder, commanded me to be silent, but nothing would stop me. However, the end was not long in coming. My words were cut off by a violent crash, which sent me spinning through space so rapidly that I couldn't speak. Down, down, down. I had one never-to-be-forgotten glimpse of that dull, level place which was to become my universe again. Then there was darkness and one final crash, and the next thing I knew I was a common, creeping square again, sitting at home in my study and listening to the peace cry of my approaching wife.

When she saw what an agitated state I was in, she insisted that I was ill and needed rest. For myself, I was glad to have an excuse to be alone in my room and think over all that had happened. As soon as I was by myself I began to feel drowsy, but before I fell asleep I tried to imagine the third dimension and especially the way a cube is formed by a moving square. It was not as clear as I wanted it to be, but I remembered the words, "upward, but not northward," and I made up my mind to remember them always.

I awoke the next morning in a joyful mood, ready to tell everyone in Flatland about my exciting journey. Just as I was thinking of how to begin, I heard the sound of many voices in the street. Then they were silent, and a louder voice was heard. It was a herald, proclaiming the order of the Council of the Circles. Anyone who was caught upsetting people with stories of revelations from another world would be committed to the state asylum.

I was taking no chances. Instead of telling people what had really happened to me, I would simply explain the truth in a way that could not fail to convince them. "Upward, but not northward" was the clue to everything. And the first person to talk to was my grandson the hexagon, whose remark about the meaning of 3^3 had already met with the sphere's approval.

I sent for the boy at once and told him that we would continue our lesson where we had left off the night before. I explained to him again how a point produces a line by moving in one dimension, and how a line produces a square by moving in two. Then, forcing a laugh, I said to him, "Now, you young scamp, yesterday you tried to make me believe that a square, by somehow moving 'upward, but not northward,' can produce some other sort of figure in three dimensions."

At that moment we heard the herald's voice again, repeating the proclamation, and young as he was, my grandson understood the situation at once.

"Grandpapa," he said, "that was only a joke. We didn't know anything about this new law last night, and I don't think I said anything about a third dimension, and I'm *certain* I didn't say one word about 'upward, but not northward.' How could something move upward and not northward? It doesn't make any sense at all."

"It makes perfectly good sense," I said. "Here, take this square for example," and I grasped a movable square that I happened to have. "I am going to move it, you see, not northward but upward somehow, that is to say, I shall move it somewhere, not like this exactly, but somehow or other . . ."

At this my grandson burst out laughing, and thinking that the lesson was nothing but a joke, he ran out of the room.

Having failed this way with my grandson, I realized that it was hopeless to try to convince anyone in Flatland that there really was a third dimension. Yet it was hard for me to keep the secret to myself. I couldn't help comparing everything I saw in two dimensions with what it would look like in three. I thought constantly about all the things I had seen during my wonderful journey to Spaceland, but as time passed it was harder and harder for me to remember just how they had looked.

At times I made dangerous statements, and once or twice I even spoke about the third and fourth dimensions. Months went by, and finally I could stand the strain no longer. Ignoring the law completely, I began telling everyone exactly how I had traveled through space with the sphere. I described everything I had seen and heard and begged people to believe me.

Needless to say, I was soon arrested and taken before the Council of the Circles.

Behind closed doors the highest circles in Flatland heard my story from beginning to end. When I had finished, their president asked me two questions:

1. Could I show him exactly which direction I meant when I used the words "upward, but not northward?"
2. Could I draw a picture of the figure that I called a cube?

Of course I could do neither, and I was committed to the asylum without further ado.

I had felt terribly alone all this time, but I needn't have. From somewhere in Spaceland, my friend had been keeping an eye on me. No sooner had the guard locked me in my room than a tiny circle appeared and widened before my eyes. It was the sphere again.

"You see," he said, "I was right. I warned you that you would be imprisoned if you tried to tell your countrymen about the third dimension, but you refused to listen to me."

"What shall I do?" I cried. "Am I never to see my family again? And all for telling nothing but the truth?"

"You'll see them," said the sphere. "And sooner than you think. I'll see to that. But I must warn you. The first year of your new era is nearly over now, and it will be nine hundred and ninety-nine years before I can come back to Flatland. If they lock you up again, I won't be back to rescue you."

"Have no fear," I told him. "I have said my last word on the subject of the third dimension. But how will you ever persuade them to free me?"

"I won't have to," said the sphere. "When the guard comes back, you will simply not be here. He won't report your disappearance, because he is the one who is responsible for keeping you here. In the meantime, you will be safely at home with your family, and you will stay there as long as you keep your secret to yourself."

With that the sphere swooped down on me and carried me off to Spaceland again. All was darkness, and the dizzy sensation came over me, but not for long. Within moments, I was safely at home, explaining to the astounded members of my family that the whole affair had been a terrible mistake.

Several years have gone by now, and I have kept the promise that I made to the sphere. Not a word about Spaceland has escaped my lips, and none ever will. I rarely go out, for fear of being seen by the high circles, who think I am still locked up in the asylum. But appearances being what they are here in Flatland, there is little danger that they would recognize me even if our paths did happen to cross.

My one remaining hope is these memoirs. I have written them on the chance that they will somehow find their way into the hands of a reader whose thinking is not confined to a world of fixed dimensions. Perhaps when he reads what has happened to me, he will be able to explore, at least in thought, all the varied and wonderful worlds of space.

VIII

The Professor's Dream:
A Visitor from Hyperspace

IT WAS LATE at night, and Professor McDougall was reading before the fire in his robe and slippers. Earlier that evening he had come across an old copy of *Flatland* in his study, and now he was unable to put it down.

"Googol," his wife called from upstairs, "are you forgetting what time it is?"

"Just a few more minutes, my dear," he answered, yawning in spite of himself.

A small flame flickered over the last few coals in the grate, and the Professor's eyelids grew heavy. Then all at once he heard a strange, hollow voice.

"Good evening Professor," it said. "Enjoying your book, I see."

The Professor glanced around the room quickly, but there was no one to be seen.

"Who are you, and how did you get into my house?" he cried.

"I'm not in your house at all," the voice answered. "I'm here in hyperspace, where there are four dimensions. Just a minute and I'll join you by the fire."

Out of nowhere a shimmering point of gold appeared before the Professor's astonished eyes. Hovering in the air before him, it blew up slowly, like a round balloon, until it was just big enough to settle itself comfortably in the chair beside him.

"A talking sphere!" cried the Professor, unable to believe his eyes.

"Not at all," came the voice again, "a hypersphere. You are seeing only my three-dimensional section. The rest of me is out here in hyperspace."

"Well, come in, come in by all means," said the Professor.

"Alas, I cannot," answered the hypersphere sadly. "Your space doesn't have enough dimensions to hold me."

"Then what do you look like?" asked the Professor. "Explain yourself."

"If you'd like," said the hypersphere, "I can pass myself through your three-space bit by bit. I'll need some more room, though. Here, let me move the sofa and the grand piano."

As the hypersphere spoke those words, the furniture disappeared into thin air.

"Don't worry," it told the Professor. "Your things will be perfectly safe here in hyperspace. Now I can let you see the rest of me."

The golden sphere moved to the place where the piano had been and began to blow up again. Bigger and bigger it grew, until it was wedged tightly between the floor and the ceiling.

"There we are," came its voice again, a little muffled this time. "It was a tight squeeze, but I made it. I'm halfway through now. From here on I'll be getting smaller."

The Professor stared as the sphere started to shrink. It grew smaller and smaller, until it was only a shimmering point again. Then it disappeared as mysteriously as it had come.

"Where are you now?" cried the Professor, who was more bewildered than ever.

"I have moved out of your space altogether."

"Do you mean upward, or off to one side?"

"Neither one. I am moving in a direction that you cannot imagine, because you are limited to a space of only three dimensions."

"But where is this direction? Can't you point it out to me?"

"If I did, you wouldn't be able to see me do it, because you don't have an eye inside of you."

"An eye inside of me! What good would that do? My inside is completely enclosed."

"For you it is," replied the hypersphere, "but from my position here in hyperspace, I can see all of you at once, your inside and your outside as well. I can see the papers in your desk drawers and the coats in your closet. And I could help myself to any one of them if I wanted to. What's that in your refrigerator? Let me see. Half a custard pie, some eggs, a cold chicken, and a bowl of jello. Orange, I believe."

"Preposterous!" cried the Professor. "I don't believe a word you're saying."

"It's easy enough to prove," said the hypersphere. "Do you think your wife can spare one of those eggs? Just wait. I won't be a minute."

The Professor scarcely had time to collect his wits when the hypersphere returned.

"Get me a small dish. There, that ash tray will do nicely. Hold it steady now."

There was a tiny splash, and the Professor was astounded to find an egg floating in the ash tray that had been completely empty a minute before.

"And here is the shell. Quite unbroken, you see. Now do you believe me? What you think of as the insides of things are completely open to me. I can take eggs out of their shells without cracking them. I can enter your house without going through the door. In fact, I can do everything in your three-dimensional world that the sphere in that book you're reading could do in Flatland. How about a gentle poke in the stomach, just to make sure you believe me?"

"Now, now," said the Professor, "there'll be no need for that. I'm quite convinced, I assure you. But tell me more about this hyperspace of yours."

"I'd be delighted to. Supppose an ant wanted to get from one point on a line to another. He could do it by moving in just one direction, so we say that a line is one-dimensional."

"Of course."

"Now the ant could get from a corner of this room to a point out on the floor somewhere by moving in two perpendicular directions. He might go five feet along the wall, say, and then three feet to the left. So the floor, being a flat surface, is called two-dimensional."

"Quite so."

"Now suppose the ant wanted to get to a point in the room that was not on the floor. The top of that lamp, for example. Starting at the corner, he could move forward along the wall until he was just opposite the lamp, then over to the left until he reached the bottom of it, and then straight up until he got to the top. Each of those three directions is perpendicular to the other two, and they are the three directions that make your space three-dimensional."

"Yes, yes," said the Professor, a little impatiently, "I know all that."

"Very well. Then you should have no trouble understanding the space I come from. In hyperspace we have a fourth dimension, perpendicular to each of your three."

"But that is impossible."

"Can you prove that it is? You are speaking on the basis of your own limited experience, just as the square was in Flatland."

"But Flatland is only a story."

"Perhaps," said the hypersphere, "but don't you realize that if there really were such a world you would be able to enter its two-dimensional houses from your third dimension just as I am able to enter your three-dimensional house from my fourth?"

"Of course," said the Professor. "There's a sphere in the book who does exactly that. But to the square in Flatland, he looks like an expanding circle."

"Just as I look to you like an expanding sphere."

"By jove, you're right!" exclaimed the Professor, who was beginning to understand things a little better now.

"Then there's the question of the egg. Can you draw me a two-dimensional egg?"

The Professor made a sketch that looked something like this:

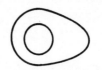

"That's fine," said the hypersphere. "Surely you can see how easy it would be for you to empty such an egg without cracking the shell. And to the Flatlander it would be sheer magic."

The Professor smiled and nodded his agreement.

"As a matter of fact, you could work a lot of magic in Flatland," the hypersphere went on. "Anything you carried off into the third dimension would be leaving the only space the Flatlanders could perceive or imagine. As far as they were concerned, it would be disappearing into thin air, just the way your sofa and grand piano did."

At that very moment the Professor's copy of *Flatland*, which had been lying on the table next to him, also disappeared.

"My book," cried the Professor, "where is my book?"

"Right here," came the answer, "not a stone's throw from where you're sitting."

The Professor craned his neck and looked all around him.

"No, not that way. In that other direction I was telling you about. The one that's perpendicular to all three of yours."

"That's amazing," said the Professor. "The book is quite invisible now as far as I'm concerned."

"I don't know what's so amazing about it. Haven't you ever lost anything before."

"Wait a minute," said the Professor. "Now that you mention it, I *have* been missing things lately. My fountain pen disappeared a short time ago, and the key to my desk drawer vanished only yesterday. And all I could find in my coat pockets this morning were two left gloves."

"You see? This is hardly a new experience for you."

"Do you mean . . . ?"

"That they might have slipped into the fourth dimension? Yes, I think it's entirely possible."

"And all this time my wife just thought I was absent-minded!"

"Nonsense. Look, I'll keep an eye out for the pen and the key. And as far as the gloves are concerned, I know I can help you there."

"How ever can you do that?"

"The same way you could do it for a Flatlander. Draw me a pair of two-dimensional gloves."

The Professor made a drawing that looked like this.

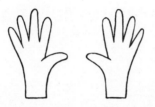

"Fine. To the Flatlander, of course, those two gloves would be entirely different. As long as they stay in the same plane, no amount of twisting and turning will make one look like the other. But you can change all that by rotating one of the gloves through your third dimension. And I can do the same thing with three-dimensional gloves by rotating one of them through my fourth. Here, let me show you with one of your slippers."

The Professor's right slipper disappeared from his foot and a left one, exactly like it, appeared on the table where the book had been. The fire had died down by this time, and the room was getting a bit chilly. The Professor, out of patience now, jumped up from his chair.

"Stop! Is there no end to your tricks?"

"Come now, Professor, it was only a joke. Let me have the left one and I'll soon set matters right again, if you'll pardon my little pun."

The Professor got back into his slippers, which were none the worse for their journey through the fourth dimension, and knotted the belt of his robe. He was still holding both ends when the knot fell apart and his robe flew open.

"Sorry, but I couldn't resist that," said the hypersphere.

"I suppose next you'll be telling me that you can untie knots without pulling the ends through them."

"But of course I can. Couldn't you do the same for a Flatlander? After all, a two-dimensional knot is just a single loop. You could untie it by turning half of it over through the third dimension, without ever touching its ends. The Flatlander would have to bring one end all the way around the other to accomplish the same thing."

The Professor made a few drawings to be sure he understood all of this.

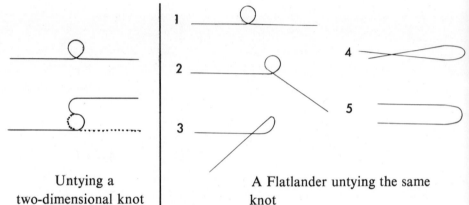

Untying a
two-dimensional knot

A Flatlander untying the same
knot

"And you can untie a three-dimensional knot just by bending part of it through the fourth dimension?"

"Certainly. Say, here's something I'd like you to see. It's called a tesseract."

A silver box appeared out of nowhere.

"That's nothing but an ordinary cube," said the Professor.

"Actually, it's a cube of cubes," answered the voice. "Don't forget, you can see only its three-dimensional section."

"And what, may I ask, is a cube of cubes?"

"I think you can understand that best by starting with a straight line, an inch long, let us say. If that line moves an inch in a direction perpendicular to itself, every point in it traces out a new line, and the result is a line of lines that you would call a square."

The Professor made a little sketch of this.

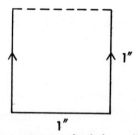

"Now let that square move an inch in a direction perpendicular to itself so that every point in it describes a line of its own. The result is a square of squares, more commonly known as a cube."

This time the Professor's sketch looked like this.

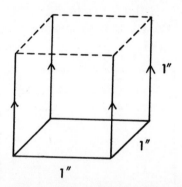

"That's as far as you can go with only three dimensions, but in hyperspace we can go one step further and move the cube in a direction perpendicular to itself. When every point of the cube, both inside and out, moves through a one-inch line all its own, so that no point passes through a place previously occupied by another, a cube of cubes is formed, and it is called a tesseract."

"But what can it possibly look like?"

"How would you tell a Flatlander what a cube looks like?"

The Professor thought for a moment.

"Well," he said, "I would tell him that when a one-inch square moves an inch in a direction he cannot imagine, it produces a three-dimensional solid that we call a cube. Each of the square's four sides forms a square as it moves, and there are two more squares in addition, one where the original square starts and one where it ends. So the cube is a portion of three-dimensional space bounded by six square faces."

"Very well," said the hypersphere, "I can describe a tesseract to you in exactly the same sort of way. When a one-inch cube moves an inch in a direction you cannot imagine, it produces a four-dimensional figure called a tesseract. Each of the cube's six faces forms a cube as it moves, and there are two more cubes in addition, one where the original cube starts and one where it ends. So a tesseract is a portion of four-dimensional space bounded by eight cubes."

All the while that the hypersphere had been speaking, the Professor had been busy sketching again.

"Look," he said, "I think I've found a way to let a Flatlander really 'see' a cube. I'd ask him to imagine a large square with a smaller one inside it and all their corners joined in pairs, like this.

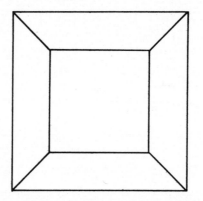

That's a cube as we'd see it if it were made of glass and we were directly above it looking down. Four of the faces are distorted, of course, because this is a two-dimensional projection of a three-dimensional figure, but if the Flatlander thought about it he would be able to count all of the cube's six faces, twelve edges and eight corners."

"Fine," said the hypersphere. "Now can you imagine a large cube with a smaller one inside it and all their corners joined in pairs?"

"Of course," said the Professor, drawing a picture as he spoke.

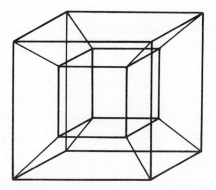

"Well," said the hypersphere, "if you built a model of that figure by gluing sticks together, you would have a three-dimensional projection of a tesseract. Six of the cubes would be distorted, but you should have no trouble finding all eight cubes, as well as twenty-four squares, thirty-two edges, and sixteen corners. Here, let me unfold the tesseract into three-space. Then you'll be able to see how it's put together."

There followed some peculiar creaking sounds, and suddenly the silver cube was surrounded by seven others exactly like it, all joined together in a sort of three-dimensional cross that stretched across the room from one end to the other.

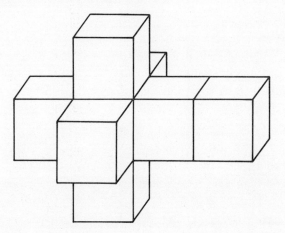

"What have you done now?" cried the Professor. "Where did all those other cubes come from?"

"I've simply unfolded the tesseract into three-space so that you can see all eight of its cubes. You could unfold a cube into Flatland in exactly the same way, you know."

"Do you mean like this?" asked the Professor, making another sketch.

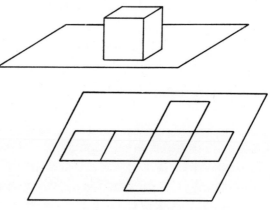

"That's right. And to the Flatlander it would seem as though six squares had suddenly appeared where only one had been before. If you could convince him that it's possible to fold those six squares up into a third dimension, he might begin to have some notion of what a cube is."

"And you can fold eight cubes into a fourth dimension to form a tesseract?"

"Precisely."

The Professor shook his head in wonder and gazed about the room. He scarcely recognized it. The sofa and the grand piano were still nowhere to be seen. On the table next to him the egg glistened in the ashtray next to its empty shell. And the tesseract filled the whole room, blocking the door completely. Just then a familiar voice was heard again, tinged this time with a note of impatience.

"Googol, what in the world has happened to you?"

"Good heavens," exclaimed the Professor, "that's my wife. She'll be down here in a minute, and I don't want her to see the place looking like this. Get that tesseract out of here, and bring back the piano and the sofa. Hurry!"

"Patience, my good man," said the hypersphere. "I'm folding the tesseract up right now."

There were a few more creaking noises, but the tesseract remained exactly where it was.

"Dear me," said the hypersphere, "this is going to be harder than I thought. The tesseract seems to be stuck."

"Well, what's to be done then?" cried the Professor. "I can't get out the door, and if my wife doesn't see me upstairs in a minute, she'll be down here directly."

"I've got it!" said the hypersphere. "Stay right where you are, and I'll have you and your wife together before you know it. I'll just bend your three-space through the fourth dimension."

"You'll do what?" cried the poor Professor, grasping the arms of his chair as hard as he could.

"Suppose you wanted to bring two Flatlanders together quickly. Wouldn't the fastest way be to bend their plane through three-space? Since they can't perceive the extra dimension, they wouldn't know the difference, and neither will you."

The Professor closed his eyes tightly and held his breath. For a moment he felt a strange sensation in the pit of his stomach, but that was all. And the next thing he heard was his wife's voice, right in his ear this time.

"Open your eyes," she said, "you must have dozed off down here. It's way past your bedtime."

The Professor looked around him. There was no sign of the tesseract. The piano was back in its usual place, and so was the sofa. The ashtray was perfectly empty.

"My book," cried the Professor. "Where is my book?"

"What book, dear?" asked his wife, helping him out of the chair.

"Why, the one I was reading all evening," he said. "It's called *Flatland*."

The Professor shook his robe. He looked under the cushions of his chair and behind the table next to it.

"Don't worry," said his wife, "you're always losing things, you know. Just last week it was your fountain pen, and yesterday you couldn't find the key to your desk. Wait and see. They'll all turn up one day just as mysteriously as they disappeared."

"Yes," said the Professor. "I expect they will."

IX

Probability and Pascal's Triangle

SUPPOSE YOU WERE at a party with, let's say, thirty other people. If someone offered to bet you five dollars that two of them had the same birthday, would you take him on?

Most people would probably jump at the chance. In fact, they might be willing to give rather high odds. With three hundred and sixty-five days to choose from, the possibility of the same one turning up twice among thirty people seems very remote.

As a matter of fact, if you made a bet like this one, the odds should be in *your* favor. Any mathematician can tell you that with only twenty-three people the chances of two or more having the same birthday are better than fifty-fifty. For more than that they're even higher. This doesn't mean that there are sure to be two people with the same birthday in any particular group of less that 366. But if the bet were made often enough in groups of twenty-three or more, in the long run there would be a tidy profit.

The solution to the birthday problem, as it has come to be called, is based on the theory of probability. Probability is a particularly interesting branch of mathematics, because it applies to so many real life situations. It doesn't require a lot of high-powered techniques either; a remarkable amount of it can be learned by simple, common-sense methods.

Here's another problem in probability that you can give your son or daughter. Assuming that boys are born as often as girls are, what are the chances of having one boy and one girl in a two-child family?

This question will probably stump your child at first, but there's a simple experiment he can do to find the answer. Tell him to list all the families he knows that have two or more children. It's important that he choose them at random. Neighbors and relatives are fine. He can use his classmates' families too, provided his school is coeducational. If it isn't, he won't be getting a random sample. All-girl families, for instance, would automatically be excluded from a boys' school.

Next to each of the names have him record the sexes of the oldest and second oldest children by writing a B or a G for each. If the Perkins family has two daughters, for instance, he would write Perkins GG. Smith BG would mean that the oldest Smith child is a boy and the next oldest is a girl.

The more families there are in the list, the more accurate the result will be. Aim for at least one hundred. When the list is finished, count the number of families with one boy and one girl. Then see what fraction this number is of the total. The longer the list is, the closer it should be to one-half. This means that if a family has two children, the chances are fifty-fifty that they have one of each sex.

A careful look at your list will show you why this is so. There are just four different combinations of letters after the names: BB, BG, GB and GG. This tree diagram is another way of finding them.

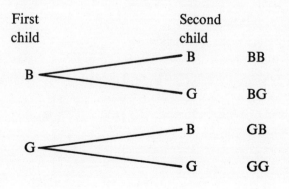

First child   Second child

B   B   BB
   G   BG
G   B   GB
   G   GG

Of the four possible combinations, two include one boy and one girl. So the chances of this particular outcome are two out of four, or one out of two. What are the chances of having two boys? Two girls?

Now consider the three-child family. What are the chances of its having, say, two girls and one boy? You could try to answer this by making another list, but it would be a long, tedious job. Suppose that instead we add another set of branches to the tree diagram.

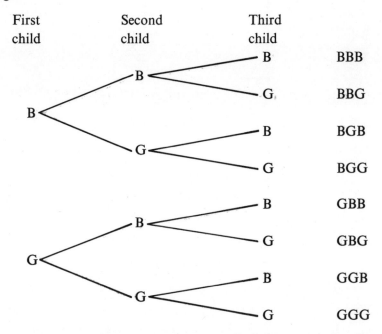

First child	Second child	Third child	

The list at the right was made by tracing each of the tree's eight branches. It tells us that there are eight different three-child families. One has three boys (BBB), one has three girls (GGG), three have two boys and one girl (BBG, BGB, and GBB) and three have two girls and one boy (BGG, GBG, and GGB). The chances of having this last combination, then, are three out of eight.

One more set of branches shows the possibilities for a four-child family.

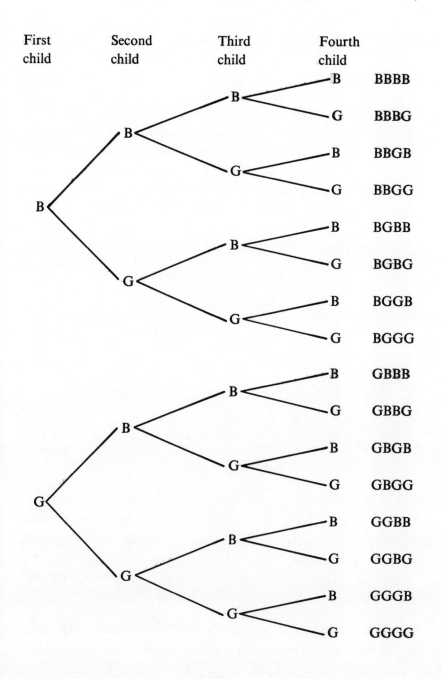

There are sixteen of these altogether. At the very top is the four-boy family, at the bottom the four-girl. The families in between fall into three different categories. Either they have three boys and one girl, three girls and one boy, or two boys and two girls. There are four in each of the first two groups and six in the third.

We could go on gathering statistics by making bigger and bigger tree diagrams, but this would be almost as tedious as making lists. With each additional child the number of branches doubles. By the time we reached the ten-child family, there would be over a thousand.

It would be much easier if there were some sort of pattern that could save us the trouble of making tree diagrams. Fortunately there is one, and it isn't hard to find it.

Let's combine the facts we already have. Here are the possibilities for the one child family,

1 boy	0 boys
0 girls	1 girl

for the two,

2 boys	1 boy	0 boys
0 girls	1 girl	2 girls

for the three,

3 boys	2 boys	1 boy	0 boys
0 girls	1 girl	2 girls	3 girls

and for the four,

4 boys	3 boys	2 boys	1 boy	0 boys
0 girls	1 girl	2 girls	3 girls	4 girls

Now in place of each of these combinations, let's write the number of times it occurs in the tree diagram.

```
              1       1
          1       2       1
       1      3       3      1
    1      4      6      4      1
```

These numbers form part of a triangle. To complete it, we need a line for the no-child family. All families with no children fall into a single category, so we put a 1 at the very top. Here is the pattern that results.

This is called Pascal's triangle, after the man who first wrote a treatise on it. Each row of the triangle depends on the row above it in a very simple way. See if your child can discover for himself that every number is the sum of the two just above it to the right and the left. (Make believe there are zeros at either end of every row.) All the 1's are just $0 + 1$. The 2 in the third row is the sum of the two 1's in the second. The 3's in the fourth row come from adding 1 and 2. To get the fifth row, we could have looked at the fourth one and said $0 + 1 = 1$, $1 + 3 = 4$, $3 + 3 = 6$, $3 + 1 = 4$, and $1 + 0 = 1$.

This gives us a way of extending the triangle as far as we like without making any more tree diagrams. The next row must be $0 + 1 = 1$, $1 + 4 = 5$, $4 + 6 = 10$, $6 + 4 = 10$, $4 + 1 = 5$, and $1 + 0 = 1$.

What does this tell us about the five-child family? The possible combinations are

5 boys	4 boys	3 boys	2 boys	1 boy	0 boys
0 girls	1 girl	2 girls	3 girls	4 girls	5 girls

and, according to Pascal's triangle, the number of times each occurs in the tree diagram is

1	5	10	10	5	1

If we had actually made the tree diagram, there would have been $1 + 5 + 10 + 10 + 5 + 1$, or 32 branches. The chances of having all girls in a five-child family, then, are 1 in 32. What are the chances of having two boys and three girls?

Pascal's triangle can help solve other kinds of problems, too. If you toss two pennies, what are the chances of one turning up heads and the other tails? This is another question that can be answered experimentally. Have your child toss two coins a number of times and record the results. If he makes enough trials, he should find that he gets one head and one tail just twice as often as he gets either two heads or two tails.

If these results sound familiar, there's a good reason for it. A tree diagram for the tossing of two coins would look like this.

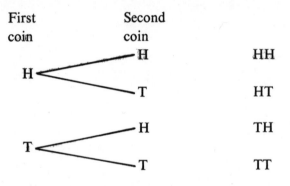

If B is substituted for H and G for T, this becomes the diagram for the two-child family. Mathematically, the two problems are identical. If there is 1 chance in 32 of having five daughters and no sons, there is 1 chance in 32 of tossing five tails and no heads. The chances of tossing two heads and three tails are 10 in 32, or 5 in 16, and these are also the chances of having two boys and three girls in a five-child family.

Pascal's triangle is much more than a collection of statistics from tree diagrams. A great many other patterns are hidden in it. Perhaps you have noticed some of them already.

Here are the first ten rows of the triangle.

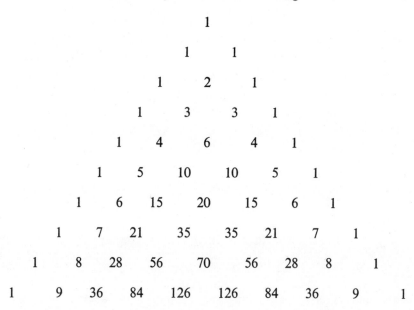

If you add the numbers in each row, you'll discover something very interesting. The sums are 1, 2, 4, 8, 16, 32, 64, 128, 256, 512, and 1024, the first ten powers of two. (The very top row has a sum of 1, which doesn't seem to follow the pattern, but 1 can be thought of as 2 to the power of zero, and this is just what we would expect it to be.)

When you think about it, these powers of two really aren't very surprising. Each time a set of branches is added to the tree diagram, the number of branches doubles. Since there are two to begin with, there are four in the next set, eight in the next, and so on.

The diagonals of the triangle have a pattern all their own. Look at the one that is marked below.

These are our old friends, the triangular numbers. What are they doing in Pascal's triangle? And why should they be in this particular diagonal?

A clue can be found in the diagonal just to the right of them. This one contains the counting numbers, 1, 2, 3, 4, 5, 6, and so on. And these are what "triangular numbers" would be in Lineland, where they could be built up in one direction only. (The 1's in the

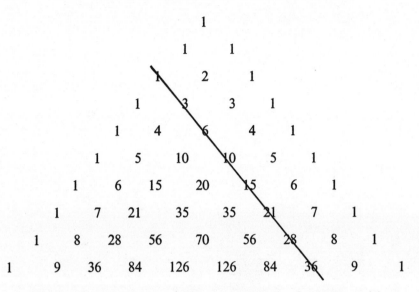

next diagonal are the "triangular numbers" of a space with no dimensions, where there is nothing but a single point.)

If the first diagonal contains the triangular numbers for no dimensions, the second for one, and the third for two, what would you expect to find in the next one? 1, 4, 10, 20, 35, 56, and 84 are the first seven tetrahedral numbers, because they represent three-dimensional triangles.

Tetrahedral numbers are built up from triangular numbers exactly the way triangular numbers are built from counting numbers. The second triangular number, 3, is formed by combining the first two counting numbers, 1 and 2, in two dimensions. The second tetrahedral number, 4, is formed by combining the first two triangular numbers, 1 and 3, in three dimensions. You can do this yourself by making a triangle from three marbles and fitting a fourth one on top of them.

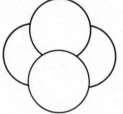

To form the next tetrahedral number, start with a triangle of six marbles and add layers of three and one.

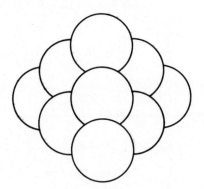

The numbers in the next diagonal, 1, 5, 15, 35, 70, and 126 are—you guessed it—the hypertetrahedral numbers. They are formed by combining tetrahedral numbers in four-space. And each successive diagonal gives the "triangular numbers" for a space of the next higher dimension.

Because of the way these numbers are manufactured from one another, Pascal's triangle is a sort of automatic adding machine. You can use it to add all the numbers in any diagonal down as far as you please. The sum will always be the number below and to the left of the last number being added.

The sum of 1, 2, 3, 4, 5, 6, 7, and 8, for example, is 36, because 36 lies below and to the left of 8. (We can confirm this with the formula we found in Chapter III.) The first seven triangular numbers, ending with 28, have a sum of 84.

Pascal's triangle has another feature that children can appreciate. Each row, if you read the digits together as an ordinary number, is a power of 11. The third row, 121, is 11 × 11. The next, 1331, is 11 × 11 × 11. Multiply 1331 by 11 and you will get the row beneath *it*, 14641.

With the next row the pattern seems to break down. 14641 × 11 is 161051, and this is not the next row of Pascal's triangle. The trouble here is that the row in question is the first to contain two digit numbers. Think of these numbers as representing powers of ten, just as the digits of an ordinary number do. 1 5 10 10 5 1 (see p. 39) would be written as $1 \times 100{,}000 + 5 \times 10{,}000 + 10 \times 1000 + 10 \times 100 + 5 \times 10 + 1$. Then the sum gives the correct result.

$$
\begin{array}{r}
100{,}000 \\
50{,}000 \\
10{,}000 \\
1{,}000 \\
50 \\
1 \\
\hline
161{,}051
\end{array}
$$

Once your child understands Pascal's triangle, he can use it to perform a very ingenious trick. Someone writes down a row of randomly chosen numbers, as many as he wishes. Your child looks at it, writes down a number of his own, and puts it aside.

Now the person builds a pyramid of numbers with his original row as a base by adding each pair of adjacent numbers and subtracting nine if the sum is nine or more. Suppose the original row was 6   5   1   8   7. He adds 6 and 5, obtaining 11, and subtracts 9 to get 2. (This type of addition is known as "casting out nines." The same result can always be gotten by adding the digits. Thus, adding the two digits in 11, 1 and 1, gives the same answer as subtracting 9 from it.) The 2 goes above the 6 and the 5, halfway between them.

```
        2
    6   5   1   8   7
```

Each pair of numbers in the row is added by this same method, and the results are placed in the row above.

```
      2   6   0   6
    6   5   1   8   7
```

When two more rows have been added, the pyramid looks like this:

```
          5   3
        8   6   6
      2   6   0   6
    6   5   1   8   7
```

At this point your child produces the number he wrote down earlier, and naturally it is the 8 that belongs at the top.

The trick is done by matching the original row with the corresponding row of Pascal's triangle, in this case 1 4 6 4 1. This tells how many times each number in the row enters into the final sum. For this pyramid, the 6 enters once, the 5 four times, the 1 six, the 8 four, and the 7 one.

With a little practice, the required number can be produced very quickly. Just think of the two rows with one above the other, like this.

$$6 \quad 5 \quad 1 \quad 8 \quad 7$$
$$1 \quad 4 \quad 6 \quad 4 \quad 1$$

Multiply each vertical pair of numbers, casting out as many nines as you can from each product.

$$6 \times 1 = 6$$
$$5 \times 4 = 2$$
$$1 \times 6 = 6$$
$$8 \times 4 = 5$$
$$7 \times 1 = 7$$

Then add the products, casting out nines again.

$$6 + 2 + 6 + 5 + 7 = 8$$

This is the 8 that belongs at the apex of the pyramid.

X

Logic and Sets

THE FIFTH GRADE was having a pet show. "How many of you are bringing dogs?" their teacher asked them. Eleven children raised their hands. "And how many are bringing cats?" Eight hands went up. "Wait a minute," said the teacher, "is anyone bringing a dog *and* a cat?" Four children said they were. How many children were bringing pets to the pet show?

Problems like this one can best be solved by thinking in terms of sets. A set is just what it sounds like, a collection of some sort. In this problem there is a set of children bringing dogs and a set of children bringing cats. The children who are bringing dogs *and* cats belong to both sets at the same time.

Sets are sometimes represented by circles. If two sets have some members in common, the circles overlap. The overlap, which is shaded in the diagram, is called the *intersection* of the two sets.

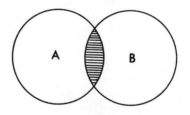

This diagram is a good one for solving our problem. Let A represent the set of children bringing dogs and B the set of children bringing cats. We know that four children are members of both sets, and these four belong in the intersection. Since eleven children are bringing dogs, there must be seven more members of set A. And since eight children are bringing cats, there are four more in set B.

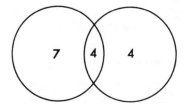

Altogether, then, there are seven plus four plus four, or fifteen children bringing pets.

Here's another problem that can be solved in much the same way. Mr. Smith was looking for a new office girl. Of eleven who applied for the job, three could neither type nor take shorthand, and four could do both. Two could type, but didn't take shorthand. How many took shorthand but couldn't type?

This time we'll use a rectangle to represent all the girls who applied for the job. Inside this rectangle there are two sets, S, the girls who took shorthand, and T, the girls who typed. And again, there is some overlap between them. The part of the rectangle that lies outside both circles represents the girls who neither typed nor took shorthand.

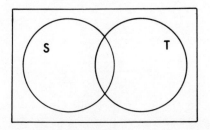

Six of the girls typed, and of these four also took shorthand. So there are six members of set T, four in the overlap between T and S and two outside it.

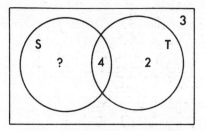

These six, plus the three who neither typed nor took shorthand, account for nine of the eleven applicants. The other two, then, must belong to the set of girls who took shorthand but couldn't type.

See if your child can figure out a diagram for this problem. After their weekly meeting, the members of the math club went out for hot dogs, and this is what they told their waitress. "None of us wants his hot dog plain. Five take mustard, six relish, and seven catsup. Four take both catsup and relish, three both catsup and mustard, and two both mustard and relish. One takes catsup, mustard, and relish." The poor waitress was thoroughly confused. "How many of you are there?" she cried. Can you help her out?

There are three sets here, so we'll need three circles, C, the set of catsup eaters, M, the set of mustard eaters, and R, the set of relish eaters.

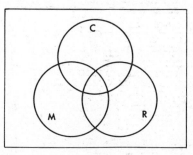

You should be able to count eight different sets in this diagram. At the very center, where all three circles overlap, is the set of people who take catsup, mustard, and relish. There are three sets who use two of the three condiments, three who use one, and, outside the three circles, one set of those who use none. One, three, three, and one make a total of eight sets.

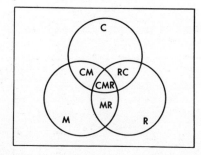

This may remind you of Pascal's triangle, and it ought to. We could have gotten the very same results by making a tree diagram. Imagine that you are offered relish, catsup, and mustard. To each you say either "yes" or "no." In how many different ways can you make your choices?

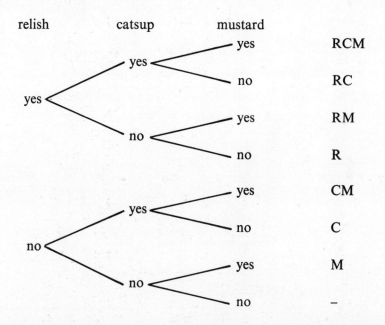

Now the problem is easy. There is only one person who takes catsup, mustard, and relish. Since four people take catsup and relish, including this one, there should be three more in the catsup-relish overlap.

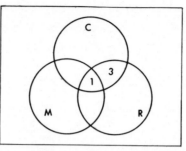

By the same token, there are two more who take catsup and mustard, and one more who takes mustard and relish.

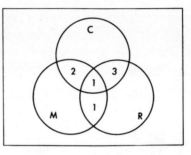

This accounts for four of the mustard eaters, five of the relish eaters, and six of the catsup eaters. There is one more member in each set.

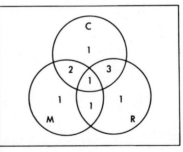

We know that no one takes his hot dog plain, so there are no members of the set outside the three circles. (A set with no members is called an empty set.) This solves the waitress's problem; she must bring a total of ten hot dogs.

When four sets are involved, four circles are needed, but this time it's more convenient to make them ovals. This diagram contains sixteen different sets. One is part of all four ovals, four are part of three, six of two, four of one, and one of none.

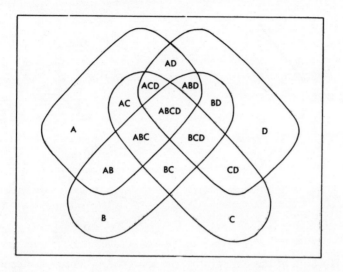

You may be reminded of Pascal's triangle again, and again there's a good reason for it.

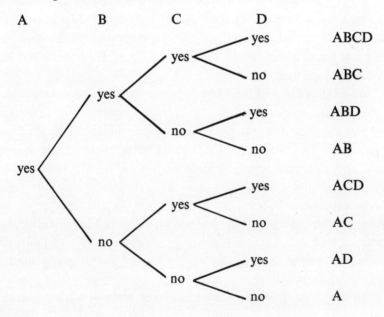

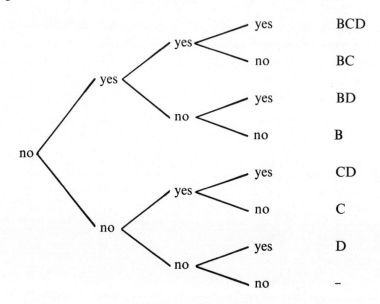

In each of these diagrams, the sets group themselves according to a row of Pascal's triangle. And the number of sets, like the sum of a row, is always a power of two.

Sixteen is two to the fourth power. And sixteen different sets can be made by combining four sets. Eight, which is two to the third power, is the number of sets that can be made from three. Does the diagram for two sets follow this same pattern? How many sets can be made from five?

When they finished their hot dogs, the members of the math club were joined by some friends, and everyone ordered a sundae. Two people had their sundaes plain. Sixteen had nuts, eleven chocolate syrup, thirteen cherries, and twelve whipped cream. Eight wanted nuts and chocolate syrup, nine nuts and whipped cream, eleven nuts and cherries, and seven chocolate syrup and whipped cream. Three ordered everything but whipped cream, two everything but nuts, one everything but cherries, and three everything but chocolate syrup. Three members of the party shot the works and had nuts, cherries, chocolate syrup, and whipped cream. How many sundaes did the waitress bring?

This problem is easiest if we start at the end and work backward.

At the center of the diagram are the three people who shot the works. Ranged about them, moving clockwise, are the one who had everything but cherries, the two who had everything but nuts, the three who had everything but whipped cream, and the three who had everything but chocolate syrup.

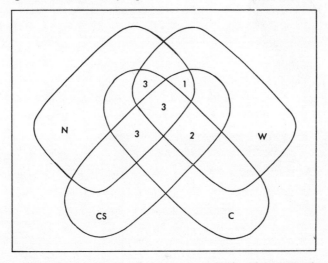

There are now a total of seven people in the nuts-chocolate syrup overlap; one more is needed. In the same way, there are two more people who take nuts and whipped cream, two more who take nuts and cherries, and one more who takes chocolate syrup and whipped cream.

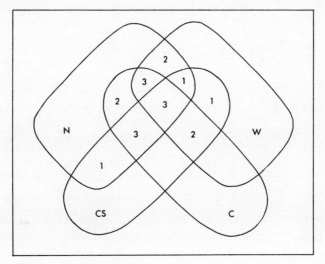

This makes fifteen nut eaters, eleven chocolate syrup eaters, thirteen cherry eaters, and twelve whipped cream eaters. We must still account for one nut eater and the two who had their sundaes plain.

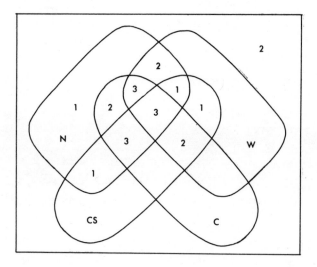

Adding all the numbers in the figure, we find that the waitress brought twenty-one sundaes.

Here's a problem adapted from one by Lewis Carroll, a mathematician who was also the author of *Alice in Wonderland*.

The Cupboard Problem

1. All the old articles in this cupboard are cracked.
2. Every jug in this cupboard is old.
3. Nothing in this cupboard that is cracked will hold water.

What conclusion can be drawn from this information?

You may be able to solve this just by thinking about it, but it can also be solved with sets. Among the things in the cupboard, there is a set of old articles, O, a set of cracked articles, C, a set of jugs, J, and a set of things that will hold water, W.

The first statement says that every old thing is a cracked thing, which means that every member of set O is also a member of set C. A diagram of these two sets looks this way:

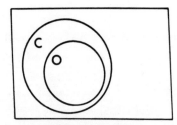

Because set O is contained entirely within set C, it is called a subset of C.

According to the second statement, J is a subset of O. This is the diagram we get when J is added to the picture.

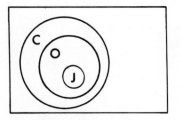

Now how about the third statement? If nothing in this cupboard that is cracked will hold water, set C has no members in common with set W, so there can be no overlap between them.

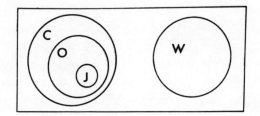

We can conclude, then, that no jug in this cupboard will hold water.

Here is another of Mr. Carroll's problems. See if your child can solve it.

The Kitten Problem

1. No kitten that loves fish is unteachable.
2. No kitten without a tail will play with a gorilla.
3. Kittens with whiskers always love fish.

4. No teachable kitten has green eyes.

5. No kittens have tails unless they have whiskers.

What conclusion can be drawn?

There are six sets in this problem:
    F: The set of kittens that love fish
   TE: The set of teachable kittens
   TA: The set of kittens with tails
    G: The set of kittens that will play with a gorilla
    W: The set of kittens with whiskers
   GR: The set of green-eyed kittens

If this rectangle represents the set of all kittens, and the circle inside it the set of teachable kittens, then the part of the rectangle that lies outside the circle is the set of unteachable kittens.

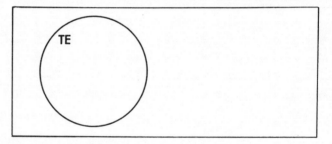

Statement 1 says that the set of kittens that love fish has no members in common with the set of unteachable kittens, and this means that it must lie entirely inside the set of teachable kittens. In other words, saying that no kitten that loves fish is unteachable is the same as saying that all kittens that love fish are teachable. Set F is a subset of set TE.

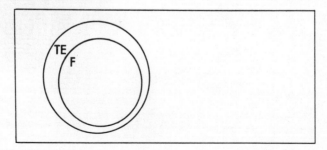

If no kitten without a tail will play with a gorilla, all the kittens that will play with a gorilla must have tails, so G is a subset of TA.

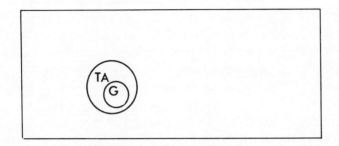

Since kittens with whiskers always love fish, W is a subset of F, and we can add it to the first diagram this way.

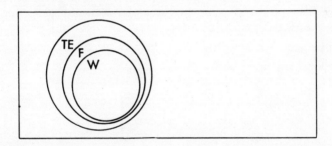

The fifth statement tells us how to combine this diagram with the second one. If no kittens have tails unless they have whiskers, all kittens with tails have whiskers and TA is a subset of W.

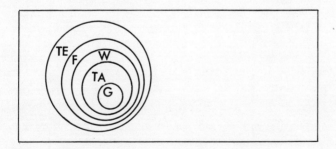

Now for the fourth statement. No teachable kitten has green eyes, so GR does not overlap TE. The final diagram looks like this,

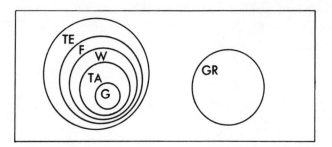

and we can conclude from it that no green-eyed kitten will play with a gorilla.

Here are some more problems you can solve with sets. The answers follow.

1.

I have thirteen aunts, and six of them are old maids. Four have blue-eyes, and seven have red hair, but only one has both, and she isn't an old maid. Only two of the old maids are redheads. Do any of the old maids have blue-eyes?

2.

What a racket there was at the breakfast table this morning! Everyone ate a different kind of cereal. Eight snapped, five crackled, six popped, and six fizzled. None of them crackled and popped, but three snapped, fizzled, and popped. In fact, all the cereals that fizzled also popped, and the only ones that crackled were the ones that snapped. How many people were at the breakfast table?

3.

There were five performers in Ricky's Combo. Three sang and three played the guitar. There were two drummers. None of the singers played the drums, but one played the guitar. There was only one announcer, and he did everything but sing. Were any of the guitar players drummers?

Here are three more problems by Lewis Carroll. What can you conclude from each set of statements?

fizzled, F. Since none of them crackled and popped, there are zeros in all four parts of the overlap between C and P. And since all the cereals that fizzled also popped, there are four more zeros in the parts of set F that do not overlap set P.

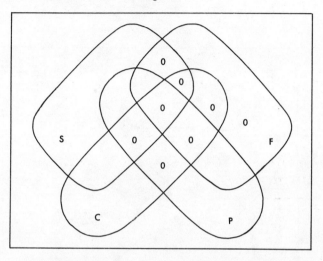

Five cereals crackled, and since the only ones that crackled were the ones that snapped, these five all go in the overlap between C and S. These, with the three that snapped, fizzled and popped, account for the eight that snapped. Six fizzled, so there must be three more in set F, and these three had to pop, too, giving us the six we need in set P.

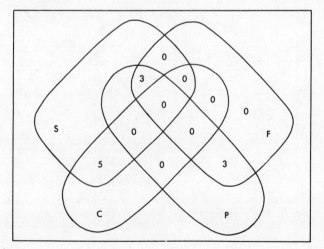

There were eleven people at the breakfast table.

3.   This is another four set problem. There are the singers, S, the drummers, D, the guitar players, G, and the announcer, A. None of the singers plays the drums, so there are four zeros in the overlap between S and D.

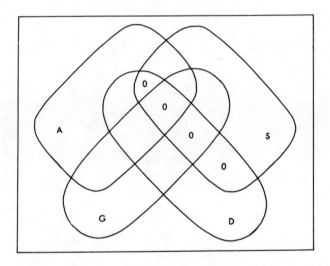

Since the announcer does everything but sing, he goes in the A,G,D overlap, and all the other sets containing A are empty.

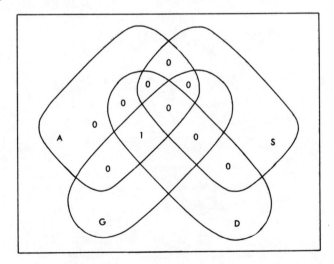

One singer plays the guitar. All the other subsets of S contain zeros, so the two remaining singers must do nothing but sing.

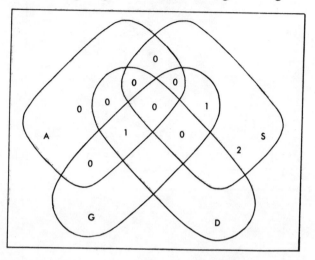

We still have to account for one guitar player and one drummer. Since there is only one more member of the combo, he must be both a guitar player and a drummer. The answer to the problem, then, is two.

4. Here the sets are R, the set of those who really appreciate Beethoven, S, the set of those who keep silence while the Moonlight Sonata is being played, G, the set of guinea pigs, and H, the set of those hopelessly ignorant of music.

From the first statement we know that no one who really appreciates Beethoven lies outside the set of those who keep silence while the Moonlight Sonata is being played. In other words, R is a subset of S.

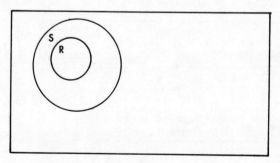

From the second statement, we know that G is a subset of H.

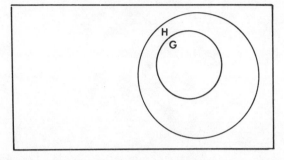

And according to the third statement, there is no overlap between set H and set S.

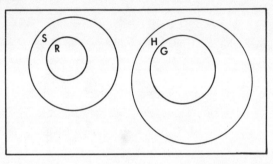

We can conclude, then, that there is no overlap between G and R; in other words, no guinea pigs really appreciate Beethoven.

5.   In this problem we must be concerned with N, the set of things that are noticed at sea, M, the set of mermaids, L, the set of things entered in the log, W, the set of things worth remembering, and I, the set of things I have met with on a voyage.

If none of the things unnoticed at sea is a mermaid, then all mermaids are noticed at sea; in other words, M is a subset of N.

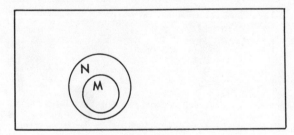

From statement 2, we know that L is a subset of W.

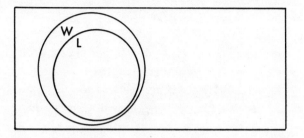

Statement 4 says that N is a subset of L, so the diagrams can be combined this way.

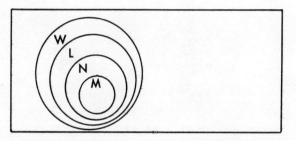

According to statement 3, there is no overlap between W and I.

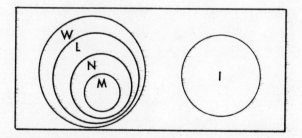

It can be concluded, then, that I have never met with a mermaid when on a voyage.

6. If you can solve this one you're really an expert. There are ten sets in this problem,

    D: the set of dated letters
    B: the set of letters that are written on blue paper

BL: the set of letters that are in black ink
 T: the set of letters that are written in the third person
 F: the set of letters I have filed
 R: the set of letters I can read
 S: the set of letters that are written on one sheet
 C: the set of letters that are crossed
BR: the set of letters written by Brown
DE: the set of letters that begin with "Dear Sir"

Statements 1, 4, 7, and 8 taken together give us the following diagram.

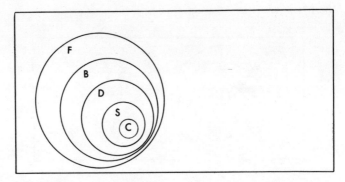

According to statement 3, there is no overlap between set F and set R, so there is no overlap between set C and set R either.

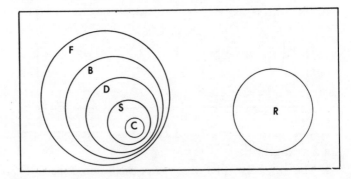

In other words, I cannot read any letters that are crossed. This means that all the letters I *can* read are not crossed. If NC is the set of letters that are not crossed, R is a subset of NC.

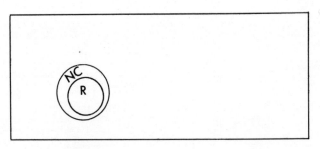

Combining this with statement 5 we have this picture,

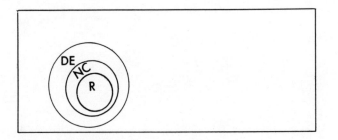

and adding statement 2 gives us this one.

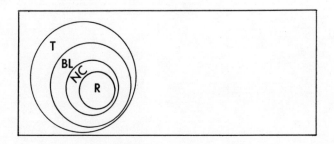

Statement 9 says that DE does not overlap T,

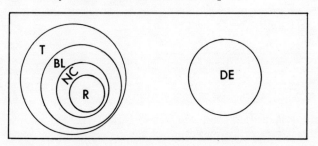

and statement 6 says that BR is a subset of DE.

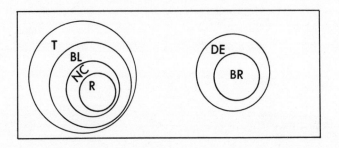

So, having considered all of the statements, we can conclude that I cannot read any letters written by Brown.

XI

A New Approach
to Geometry

THERE'S A VERY simple and pleasant way for you to explore the facts of geometry with your child, and that is by folding and cutting paper. Wax paper is best for this. Its transparency is a great convenience, and when it is creased it forms white lines that are easy to see.

Here's an easy way to show that the three angles of a triangle add up to a straight angle. First form a triangle by making three intersecting creases in the wax paper. They should be as straight as possible. Then cut the triangle out along these creases.

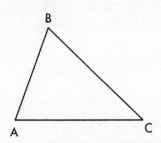

Next, you'll need to form an *altitude*. An altitude, you may remember, is a line through a vertex of a triangle perpendicular to the opposite side. A perpendicular may be formed by folding a line over on itself and creasing; the crease will be perpendicular to the line. A perpendicular can be made to pass through any point, either on the line or off it. In this case, the perpendicular must pass through a vertex of the triangle, so, fold the triangle through one vertex, lining up the edges of the opposite side as carefully as you can.

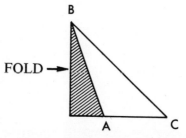

Then crease along the fold, and open the triangle out again.

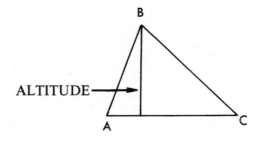

Now fold the altitude over on itself so that the vertex just touches the opposite side. This is where the transparency of the wax paper is so helpful; if you hold the triangle up to the light you can make sure that the two halves of the altitude coincide exactly.

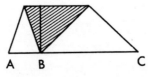

Fold the other two angles over to meet the third one; if you've worked carefully, they should fit exactly. The triangle's three angles form a perfect straight angle.

A, B
and C

Any triangle can be turned into a rectangle by this procedure. If your child knows how to find the area of a rectangle, he should be able to find the area of a triangle too.

Suppose the base of a rectangle is 7 units and the height is 4. If these units are marked off and lines are drawn through them parallel to the rectangle's sides, 28 little squares are formed, and each of them is one unit on a side.

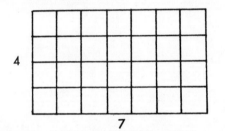

This rectangle, then, has an area of 28 square units. In general, the number of square units in any rectangle is found by multiplying the base by the height. (Actually, this argument is fudging a little, because it assumes that a unit can be found that will measure both the base and the height. This isn't always possible, but the formula is valid anyway.)

How does the folded rectangle compare in area with the triangle it was made from? Since its thickness is double, there are really two rectangles, and each of them is half the original triangle. Their base is half the triangle's base; their height is half its height. So if b is the triangle's base and h is its height, each rectangle has an area of $\frac{1}{2}b \times \frac{1}{2}h$, or $\frac{1}{4}bh$. Multiply this by two, and you get the familiar formula for the area of a triangle: $A = \frac{1}{2}bh$.

Geometry students are always surprised to learn that the three altitudes of a triangle pass through a common point, and when you think about it, it *is* rather remarkable. Two lines, of course, always have a point in common unless they are parallel. But when a third line also shares this point, the situation is a very special one.

Paper folding is an excellent way of showing that this special situation exists for the altitudes of a triangle. Cut out a fairly large triangle, and carefully fold in all three of its altitudes, one through each vertex. No matter what the triangle's shape, the altitudes will all meet at a point inside it.

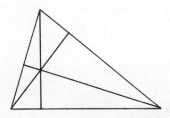

Altitudes aren't the only set of lines in a triangle that have a common point. *Angle bisectors*, *perpendicular bisectors*, and *medians* have this property, too, and all of them can be made by folding paper.

To bisect an angle, line its sides up carefully and make a fold through its vertex.

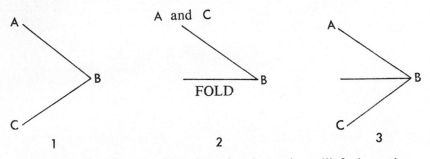

Do this to the three angles of a triangle, and you'll find another common point.

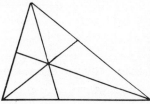

The point where the angle bisectors meet is the center of a very special circle, one that just fits inside the triangle. You can find this circle's radius by folding a line perpendicular to one of the triangle's sides and passing through the point where the bisectors meet. The line segment between this point and the side of the triangle is the circle's radius. Set a compass at this length, and draw in the circle.

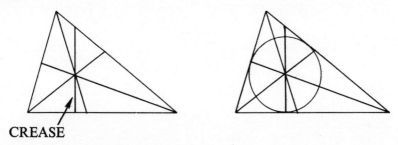

CREASE

Another circle has its center at the point where the perpendicular bisectors of the triangle's sides meet. This circle lies outside the triangle and passes through all three of its vertices.

The perpendicular bisector of a line is a line which is perpendicular to it at its midpoint. To fold the perpendicular bisector of a side of a triangle, bring its endpoints together, line up the edges between them, and crease.

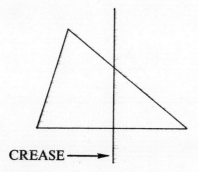

CREASE ⟶

The three perpendicular bisectors will have a common point, although it may not lie inside the triangle.

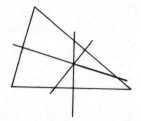

This point is the same distance from each of the three vertices, and this distance is the radius of a circle, centered at the point, that just encloses the triangle.

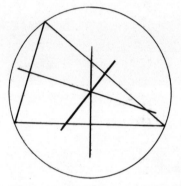

A median is a line that goes from a vertex of a triangle to the midpoint of the opposite side. To find the midpoint of a side, bring its endpoints together and make a crease halfway between them. Then form the median by making a fold through this midpoint and the vertex opposite it.

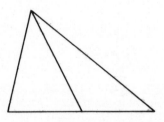

The point where a triangle's medians meet is its balancing point. Theoretically, if this point were supported by the head of a pin, the triangle should remain perfectly horizontal. This may not work for a wax paper triangle because it is subject to air resistance, but if you transfer the triangle to a piece of cardboard and try balancing it at this point you should get better results.

It's easy to fold parallel lines if you make use of the fact that two lines that are perpendicular to the same line are parallel to each other. Fold perpendiculars to a line at two different points, like *AB* and *CD* in the picture below.

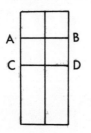

These lines will be parallel to each other.

If another pair of parallel lines are folded across these two, a parallelogram is formed. When the two sets of parallel lines are perpendicular to each other, the parellelogram is a rectangle.

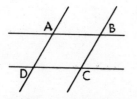   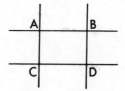

ABCD is a parallelogram,   ABCD is a rectangle
   but not a rectangle

It should be interesting to see how many of the parallelogram's properties your child can discover for himself. Encourage him to look for equal lines and equal angles. He can test for these by superimposing them to see if they coincide. Here are some of the discoveries he may make:

1. The opposite sides of the parallelogram are equal.
2. The opposite angles are equal too.

3. If the two diagonals are folded in, they meet at a point that divides each of them in half.

4. Each diagonal divides the parallelogram into two triangles that will coincide if one of them is cut out and turned over. The two diagonals together divide it into two such pairs.

5. In the case of the rectangle, the diagonals are equal.

A rhombus is a parallelogram with four equal sides. To make a rhombus, fold a pair of parallel lines, and fold a third line cutting both of them, but not at right angles.

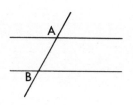

The segment that the parallels cut off on this third line (*AB* in the picture) will be one side of the rhombus. Find a point *D* on the upper parallel so that *AD* equals *AB* and a point *C* on the lower parallel so that *BC* equals *AB* too.

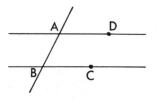

Then fold a line through *D* and *C*. This makes *ABCD* a rhombus.

The rhombus has several special properties. For one thing, its diagonals are perpendicular. This means that the four triangles they form are exactly the same size and shape. Cut them out, and you'll find that you can pile them up with all their edges coinciding.

Encourage your child to see what other discoveries he can make about parallelograms. You might suggest that he fold in the bisectors of all four angles. If these bisectors are long enough, they'll meet to form a rectangle. (The rectangle won't lie inside the parallelogram unless one side of the parallelogram is less than twice the length of the other.)

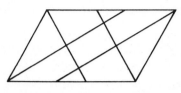

If the parallelogram is a rhombus, this rectangle disappears. See if your child can discover why.

Once you have formulas for the area of a rectangle and a triangle, it's easy to work out one for the parallelogram. There are at least two ways to go about doing this. If an altitude is folded through one vertex of a parallelogram, the triangle that is formed can be cut off and put back on the other side.

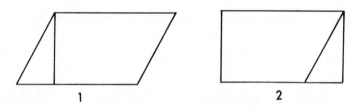

This turns the parallelogram into a rectangle with exactly the same area. And since the rectangle's area is its base times its height, the parallelogram's is too.

Another way to find the area of a parallelogram is to use a diagonal to divide it into two triangles.

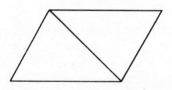

Each triangle has an area of $\frac{1}{2}bh$, so the two together must be $2 \times \frac{1}{2}bh$, or just bh.

Circles, too, can be studied by folding paper. To find the center of a circle, fold in two of its chords. (A *chord* is a line that connects two points on the circle's circumference.)

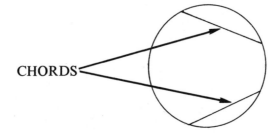

Now fold the perpendicular bisectors of the chords.

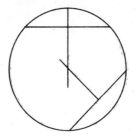

The point where they meet is the center of the circle.

Once your child knows how to find the center of a circle he can make a number of other discoveries. Here are a few of them:

1. A line that bisects the angle between two radii also goes through the center of the chord that joins the ends of the radii.

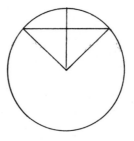

2. A diameter that goes through the center of a chord is also perpendicular to it.

3. If two chords are perpendicular to the same diameter, the parts of the circle that lie between them are equal.

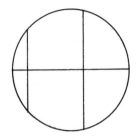

4. Lines that go from a point on a circle to the ends of a diameter are perpendicular to each other.

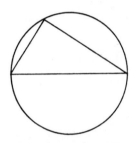

Here's a paper folding exercise that takes the form of a puzzle. It's a demonstration of a rather astonishing fact about right triangles that was proven 2500 years ago by a Greek named Pythagoras. A right triangle, of course, is one that has a right angle. The sides that form this right angle are called the *legs*; the third side is the *hypotenuse*. The Pythagorean theorem, as it is usually taught in school, says that the sum of the squares of the two legs is always equal to the square of the hypotenuse.

The use of the word "of" suggests that the sides of the triangles are thought of as numbers rather then lengths. The square of a number is the number that you get when you multiply it by itself. If, for example, the legs were three inches and four inches long, the square of the hypotenuse would be 3 × 3 plus 4 × 4, or 25. The hypotenuse would then be five inches long.

Actually, this numerical approach was not exactly what Pythagoras had in mind. He spoke of squares *on* the triangle's sides, and what he meant was a construction like this.

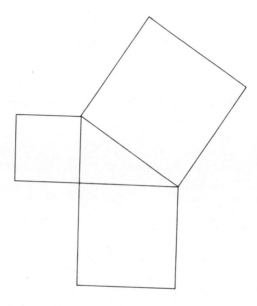

The areas of the two smaller squares taken together, he said, were just equal to the area of the largest one.

There's a very simple way to demonstrate this by folding and cutting paper. To form the right triangle, fold two perpendicular lines and a third line passing through both of them.

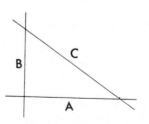

Now fold squares on both legs of the triangle. To do this, fold the side labeled *a* in the drawing over to the extension of *b* that is next to it, and mark off on it a segment equal in length to *a*.

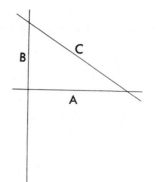

SEGMENT EQUAL TO A

Through the mark, fold a line perpendicular to the extension of *b*, and mark off another equal segment. Then fold in the fourth side of the square. Repeat the procedure to form a square on the other leg.

Now you're ready to show that Pythagoras was right. Fold both the diagonals of the larger square; they will meet at its center. Through this center, fold a line parallel to the triangle's hypotenuse. Then fold another line through the center perpendicular to this one. These two lines will divide the square into four identical parts.

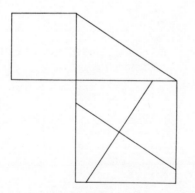

Cut them out, along with the square on the smaller leg.

Now see if your child can put these five pieces together to form a square on the hypotenuse. He may find it easier to do this if you transfer the pieces to heavy paper or cardboard. Not fair turning the page until you discover the solution for yourself!

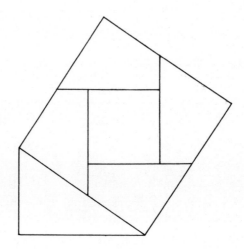

XII

The Conic Sections

THE WORLD AROUND us is filled with mathematical curves. Among the most common are a group associated with one of the simplest of all solids, the cone.

Throw a ball upward at an angle, and it will follow a curved path called a *parabola*.

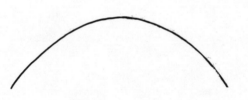

Huge telescopes like the one at the Palomar Mountain Observatory of the California Institute of Technology have parabolic mirrors. Their shape makes it possible for them to gather light from distant sources and focus it sharply at a point.

You can make a parabola by slicing an ordinary ice cream cone with a sharp knife, parallel to its edge.

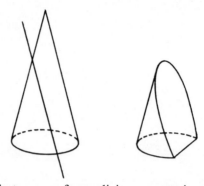

Another curve that comes from slicing a cone is an oval called an *ellipse*.

To make one, slice the cone on a slant.

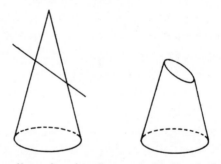

The whispering gallery in the Statuary Hall of the Capitol in Washington, D.C., has an ellipsoidal ceiling. A whisper originating at a point near one end can be heard clearly at a point near the other.

A *hyperbola* is a curve with two branches.

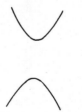

It is made by slicing through two cones held end to end.

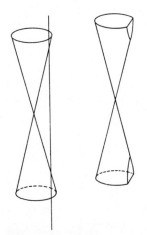

Some comets follow hyperbolic paths.

This family of curves, the parabola, the ellipse, and the hyperbola, are called the *conic sections* because of their relationship to the cone. The ancient Greeks studied them nearly 2500 years ago, but during the two thousand years that followed they were nothing but mathematical curiosities. Then, early in the seventeenth century, Johannes Kepler discovered that all the planets travel in elliptical orbits with the sun at a special point inside. Before the seventeenth century was over, Sir Isaac Newton was able to show that the force of gravity produced these elliptical orbits, and the conic sections took their place among the most important curves in the physical world.

Today the conics are a standard part of the high school math curriculum. Their equations are derived from their definitions, and their properties are thoroughly explored. This kind of study requires a knowledge of both algebra and geometry, and most students aren't ready for it until they reach the eleventh or twelfth grade. But long before this, children can learn a surprising amount about the conic sections by constructing them in various ways.

Loop a string around two thumbtacks set into a board some distance apart, and tie both ends of the string to a pencil. Keeping the string taut, move the pencil all the way around the tacks. As you do, you will trace out the oval path of an ellipse.

Suppose the string that forms the loop is 12″ long and the tacks are 4″ apart. Then the length of the string stretched between the two tacks and the pencil will always be 8″. If the pencil is 6″ from one tack, it will be 2″ from the other. If it is 5″ from one, it will be 3″ from the other. The pencil's path can be thought of as the set of points whose distances from the two tacks always add up to 8″. All ellipses can be defined this same way, as sets of points whose distances from two fixed points add up to a constant.

The hyperbola is a close relative of the ellipse. For the points on an ellipse, the *sum* of the distances from two fixed points is a constant; for the points on a hyperbola, the *difference* between these distances is a constant.

You can make a hyperbola with a ruler, a piece of string, and a pencil. Be sure the string is shorter than the ruler. Mark two points on a piece of paper so that the distance between them is equal to the difference between the lengths of the ruler and the string. Attach one end of the string to one of these points and one end of the ruler to the other. Then attach the free end of the string to the free end of the ruler.

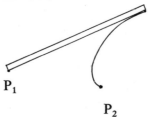

P_1

P_2

Using the point of a pencil to hold the string taut against the side of the ruler, rotate the ruler around the point at which it is attached. The pencil, moving along the taut string and ruler, will trace out half of one branch of a hyperbola.

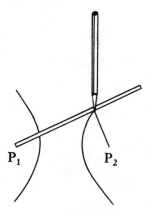

Why is this curve a hyperbola? Suppose the ruler is 12″ long and the string is 9″. If the pencil is holding an inch of string against the ruler, it is 11″ from one of the points and 8″ from the other. The difference between these distances is 3″. If four inches of string are held against the ruler, the pencil's distances from the points are 8″ and 5″. Again the difference is 3″, and it will remain so throughout the pencil's path. Because this difference is a constant, the points on the curve satisfy the definition of a hyperbola.

The third conic section, the parabola, is a set of points that are just as far from a fixed point as they are from a fixed line. A parabola can be constructed with a piece of string and a right triangle made of plastic or cardboard. The string must be the same length as one of the triangle's legs. Attach one end of the string to the point where this leg meets the hypotenuse, and line up the other leg with a straight line drawn on a piece of paper. Attach the free end of the string to a point near this line.

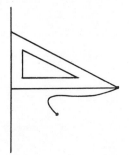

Now, using a pencil to hold the string taut against the side of the triangle, slide the triangle up and down the line. As you do, the pencil will always be just as far from the line as it is from the point, and it will trace out a parabola.

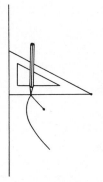

The conic sections can also be made by folding paper. To make an ellipse, cut a circle out of wax paper. The larger it is, the better. Choose a point on the circle (not the center), and mark it clearly. Then lay the circle flat on a table, bring a point on its circumference up to meet the chosen point, and crease.

CREASE

THE CHOSEN POINT

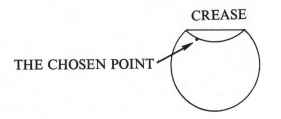

Do this twenty or thirty times, working your way around the circle. When you're finished, the lines will enclose an ellipse with one of its fixed points at the center of the circle and the other at your chosen point. The more creases you make, the smoother the ellipse will be.

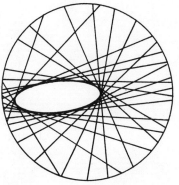

To fold a hyperbola, draw a circle on a piece of wax paper, but this time don't cut it out. Instead, mark a point outside it, fold this point onto the circle's circumference, and crease.

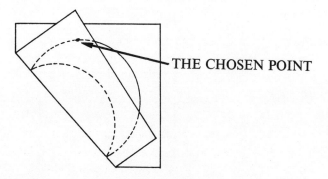

THE CHOSEN POINT

Repeat this procedure again and again, working your way around the circle as you did for the ellipse. When you are finished, the two branches of a hyperbola will be outlined by the creases.

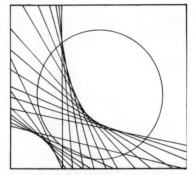

To make a parabola by paper folding, start with a straight line and a point about three inches away from it. Fold the point onto the line and crease.

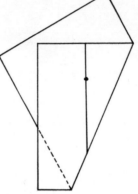

Continue doing this up and down the line until the creases envelop a parabola.

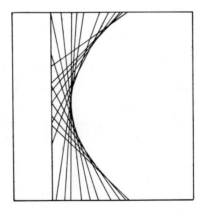

The proofs for these constructions depend only on elementary geometry, but you may find them a little involved. Skip them if you'd like to.

In the case of the ellipse, each crease is the perpendicular bisector of a line joining the chosen point, *C*, to *P*, a point on the circle's circumference.

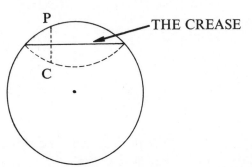

This means that every point on the crease is just as far from one of these points as it is from the other.

In the picture below, *Q* is the point where the radius drawn to *P* crosses the crease.

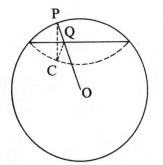

Since *Q* is on the crease, we know that *QP* equals *QC*. This means that *OQ* + *QP* is exactly the same as *OQ* + *QC*, and this will be true no matter where *P* is. But the line going from *O* through *Q* to *P* is a radius of the circle; its length never changes. So the sum of *OQ* and *QC* will always be a constant too. This makes *Q*, by definition, a point on an ellipse with one fixed point at *C* and the other at *O*.

For any other point on the crease, say *R*, $OR + RP$ is greater than $OQ + QP$, because *OQP*, being a straight line, is the shortest distance from *O* to *P*.

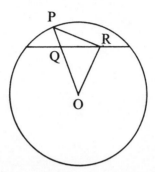

This means that *Q* is the *only* point on the crease that lies on the ellipse; all the others lie outside it. A line like this crease, that touches a curve at only one point, is called a *tangent*. Every crease contains just one point like *Q*, so every crease is a tangent to the ellipse. These tangents, if there are enough of them, will appear to envelop the curve.

The construction for the hyperbola works in much the same way as the one for the ellipse. Every crease is the perpendicular bisector of a line joining a point on the circle to the chosen point outside it.

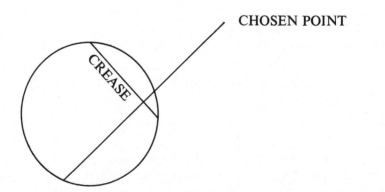

In the diagram below, *C* is the chosen point, *P* is a point on the circle, and *O* is the center of the circle. A diameter has been drawn through *P*, and it crosses the crease at *Q*.

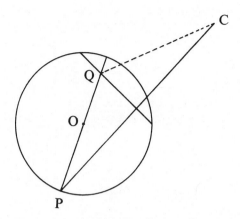

Since Q lies on the crease, it is just as far from C as it is from P, and $QP = QC$. Then $QP - QO$ is the same as $QC - QO$. But $QP - QO$ is OP, the circle's radius, and it is a constant. This makes $QC - QO$ a constant, too, and according to the definition, places Q on a hyperbola with one fixed point at C and the other at O. Since every other point on the crease lies outside this hyperbola, the crease is a tangent to the curve.

The paper-folding construction for the parabola is based on the same property that makes it such a useful curve in the design of telescopes and other optical instruments. All the points on a parabola are just as far from a fixed line as they are from a fixed point. If the parabola is a mirror, any ray of light traveling perpendicular to the fixed line will be reflected through the fixed point. Such a ray is shown in the picture below. According to the law of reflection, the angles that it makes with the tangent to the curve at R, the point of reflection, are equal.

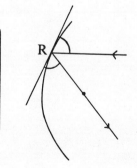

Here the incoming ray is extended to meet the fixed line at D, and *D* is joined to the fixed point, *F*. Since *R* is a point on the parabola, its distance from the fixed line, *RD*, is equal to its distance from the fixed point, *RF*, so triangle *RDF* is isosceles. And since the opposite angles formed by intersecting lines are equal, the tangent to the parabola at *R* bisects the vertex angle of this isosceles triangle. This makes it the perpendicular bisector of the base, *DF*.

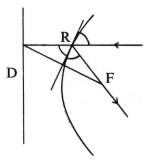

But *DF* is a line joining the fixed point to the fixed line. Since the creases were made by folding the point onto the line, each crease is the perpendicular bisector of a line segment like *DF*. And since a line segment can have only one perpendicular bisector, the crease and the tangent must be one and the same.

XIII

Using Mathematics

Now MORE THAN ever before, mathematics is the language of science, and for many children this is one of its most appealing aspects. A number of simple experiments can be performed with very ordinary objects, and children do many more of these than they realize. To balance a seesaw, for instance, the larger child soon learns that he must sit closer to the center than the smaller one. But how much closer? Can the balancing point be predicted?

To answer these questions, all you need are two children, a scale, a seesaw, and a tape measure. Weigh the children carefully before you start. Then let them adjust themselves until the seesaw is balanced, and record the distance of each child from the center. Your results should look something like this:

	John (40 lbs.)	Jane (60 lbs.)
distance from center	4½ ft.	3 ft.

Now let the larger child move a foot closer to the center, and see if the smaller one can find another balancing position. This will give you a second set of data:

	3 ft.	2 ft.

Is a pattern emerging here? Try balancing yourself against both children. If you weigh 150 pounds and the combined weight of the children is 100, when you are 4 feet from the center, they must be 6. You may have noticed that every time a balance is achieved, the product of the weight at one end of the seesaw and its distance from the center is equal to the product of the weight at the other end and *its* distance from the center. In this case, $150 \times 4 = 100 \times 6$.

Once you have learned this, you can predict new balancing positions. Suppose you are 2 ft. from the center. Where must the 60-pound child be to balance you? 150 times 2 equals 60 times what? Is your prediction accurate?

The pendulum also lends itself well to a mathematical description. Tie a fairly heavy object to a length of clothesline, or a piece of twine, and hang it from an overhead hook or an exposed pipe. Start off with the weight just a few inches from the floor, and measure the length of the rope as carefully as possible. You'll be taking this measurement many times, and it's important that you have a standard procedure for doing it.

Now pull the weight a short distance away from its resting position, and let go of it so that it swings back and forth from one side to the other. The time that it takes to make a round trip, from one side to the other and back again, is called the *period*. It is an interesting property of pendulums that their period depends only on their length and not on the distance covered in the side-to-side swing. (This is the basis for their use in clocks.) The object of this experiment is to see how the period of a pendulum is related to its length.

A stopwatch is especially convenient for measuring the period, but if you don't have one, a clock with a second hand will do. It's very difficult to time a single oscillation accurately, so time ten of them and take a tenth of the result. If you're using a clock, you and your child will have to work as a team. One person says "start" as he releases the weight; the other notes the time to the nearest second. (If the clock is big enough, it may even be possible to estimate tenths of a second.) The first person counts ten round trips and says "stop"; the other notes the time again. One tenth of the interval that elapsed is the period of the pendulum for this particular length.

Now shorten the pendulum a few inches, and measure the period again. You'll find that it's not quite as long as it was before. Continue shortening the pendulum, a few inches at a time, and measuring the period for each new length. Keep a careful record of the results.

Here are some measurements that were taken with a brick tied to a piece of clothesline and hung from an overhead pipe.

length of rope	time for ten oscillations	period
$64\frac{3}{4}''$	27.2 sec.	2.72 sec.
$62\frac{1}{4}''$	25.8 sec.	2.58 sec.
$51\frac{1}{2}''$	23.7 sec.	2.37 sec.
$38''$	20.6 sec.	2.06 sec.
$31\frac{1}{2}''$	18.7 sec.	1.87 sec.
$27\frac{1}{2}''$	17.5 sec.	1.75 sec.
$26''$	17.1 sec.	1.71 sec.
$24''$	16.6 sec.	1.66 sec.
$22\frac{1}{2}''$	16.1 sec.	1.61 sec.
$18\frac{3}{4}''$	14.9 sec.	1.49 sec.
$15\frac{3}{4}''$	13.8 sec.	1.38 sec.
$12''$	12.0 sec.	1.20 sec.
$10''$	11.9 sec.	1.19 sec.
$4\frac{3}{8}''$	9.0 sec.	.90 sec.
$2\frac{1}{4}''$	8.0 sec.	.80 sec.

The best way to see how the period is changing with the length is to make a graph of these results. (You can buy graph paper in most stationery stores.) Use a vertical line for the length and a horizontal line for the period. These lines are called *axes*. The zeros on both axes should coincide at a point in the lower left hand corner of the graph.

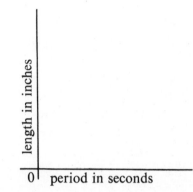

Now you must decide how many squares will represent one inch on the vertical axis and how many will represent one second on the horizontal. This is called choosing a scale. The larger the scale is, the larger the graph will be, and the easier it will be to read. On the other hand, the scale must be small enough to permit all the data to be recorded.

On the graph shown here, each square on the vertical axis represents one inch, and each square on the horizontal axis represents one-twentieth of a second. To plot the lowest pair of values (length $2\frac{1}{4}''$, period .8 seconds), start at the zero and move sixteen squares to the right on the horizontal axis. Since $\frac{16}{20}$ equals $\frac{8}{10}$, the vertical line immediately to the right of the sixteenth square is the line on which the time is .8 seconds. Now move up along this line $2\frac{1}{4}$ squares. (The fourth of a square must be estimated.) This should bring you to the first point on the graph.

As more and more points are plotted, a curve begins to appear. Every point may not lie right on it, but it should be possible to draw a smooth curve with the points distributed evenly on either side of it. If the pendulum's period were timed for every possible length, all the pairs of values would lie in the neighborhood of this curve.

Now we have a way of predicting the period for lengths we haven't tried yet. How long would the period be if the length were 45 inches?

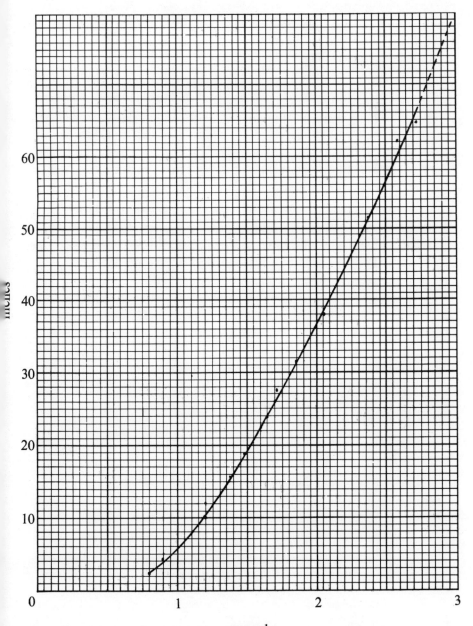

seconds

To find out, move up the vertical axis to 45, and out along the horizontal until you reach the curve. Then move straight down to the horizontal axis and read off the period that corresponds to a length of 45″. It's a little more than 2.2 seconds, and this can be verified experimentally.

The curve can even give us information about experiments we can't perform. Suppose the ceiling had been higher and it had been possible to make the pendulum 80 inches long. What would the period have been then? We can find out by drawing an extension of the curve, shown by the dotted line at the top of the graph. At 80 inches, it appears, the period would have been somewhere in the neighborhood of 3 seconds.

The results of this kind of thinking are all around us. The world population in ten years' time can be predicted with surprising accuracy. Returning astronauts are met promptly by a rescue vessel; their splashdown point is computed well in advance by observers many miles away. The child who has seen mathematics applied to the physical world can begin to understand how these things are possible.

XIV

Some Famous Unsolved Problems

Too many people think of mathematics as a dusty collection of well-established facts. In reality, it is one of the most exciting intellectual disciplines, very much alive and growing all the time. New discoveries are constantly being made, and a number of important problems remain to be solved.

Not all of these problems are practical ones. Many arise from simple curiosity. Their solutions would have no practical applications, at least none that anyone can see in advance. But this makes no difference to the mathematician. He values solutions for their own sake, whether they happen to be useful or not.

The *prime numbers* have been a source of fascination for many centuries. These are numbers like 17 and 31, that have no divisors but themselves and one. You can discover all the primes under 100 by writing out the numbers from 1 to 99 and eliminating all that are *not* prime, that is, all that are multiples of smaller numbers. 2 is prime, but none of its multiples are, and this rules out all the other even numbers. 3 is prime too, but every third number after 3, being a multiple of 3, must be eliminated. (Half of these are multiples of 2 as well, and will have been crossed out already.) The next prime is 5, so every fifth number after 5 goes out. So does every seventh number after 7. It shouldn't take you long to discover that there are just 25 primes under 100, not counting 1.

This procedure is called the Sieve of Eratosthenes, after the man who devised it two thousand years ago. It's rather tedious, but it works. High speed computers can use it to discover enormous primes like 96,079 and 1,111,111,111,111,111,111, But mathematicians aren't satisfied just to identify the primes one by one. They would be much happier if they could discover some sort of formula that would turn out primes automatically. Here is an example of such a formula.

$$1 \times 1 + 1 + 41 = 43$$
$$2 \times 2 + 2 + 41 = 47$$
$$3 \times 3 + 3 + 41 = 53$$
$$4 \times 4 + 4 + 41 = 61$$
$$5 \times 5 + 5 + 41 = 71$$
$$6 \times 6 + 6 + 41 = 83$$

The six numbers on the right are among the primes you have already found. $7 \times 7 + 7 + 41$ is a prime too. So is $23 \times 23 + 23 + 41$. As a matter of fact, all the numbers from 1 to 40 will give primes when they are substituted for N in the formula $N \times N + N + 41$. With 41, though, the system breaks down. $41 \times 41 + 41 + 41$ is 1763, which is 43×41 and clearly not a prime.

So far no formula has been found that continues to manufacture primes indefinitely. It may very well be that there isn't any. But until there is decisive proof one way or the other, the search will continue.

Another famous problem began with a tantalizing note that Pierre de Fermat, a seventeenth-century French mathematician, wrote in the margin of a book more than three hundred years ago. "To divide a cube into two cubes, a fourth power, or in general any power whatever, into two powers of the same denomination above the second is impossible, and I have assuredly found an admirable proof of this, but the margin is too narrow to contain it."

Every high school student should know that $3^2 + 4^2 = 5^2$. 3, 4, and 5 are a so-called Pythagorean triple; they can represent the sides of a right triangle because they satisfy the Pythagorean theorem. So do 5, 12, and 13; 7, 24, and 25; and 8, 15, and 17. You can easily demonstrate that

$$5^2 + 12^2 = 13^2$$
$$7^2 + 24^2 = 25^2$$
$$\text{and } 8^2 + 15^2 = 17^2.$$

Fermat's contention was that no such equations can be found for powers higher than the second, that is, that there are no positive whole numbers that satisfy the equation $x^n + y^n = z^n$ when n is greater than 2.

Subsequent mathematicians have succeeded in showing that this is true for all values of n less than 600, and for a great many others as well. But so far no one has been able to prove that it is true for any number whatever.

Did Fermat really have such a proof? No one will ever know. But until somebody manages to discover it all over again, or to prove that it can't be done, this question too will continue to interest mathematicians the world over.

Another famous unsolved problem has to do with map coloring. What is the least number of colors you can use to color a map if regions with a common border are always colored differently?

The best way to understand this problem is to color an actual map. Let's start with the western part of the United States. Three different colors will be needed for Washington, Oregon, and Idaho, because each shares a border with the other two. Suppose we make Washington red, Oregon green, and Idaho blue.

Then Nevada, which shares borders with Oregon and Idaho, must be red, too, and California must be blue.

Now we are forced to color Arizona green,

and when we come to Utah, which shares borders with Idaho (blue), Nevada (red), and Arizona (green) we need a fourth color. Let's make it yellow.

Will these four colors be enough to finish the map? Or will a fifth be necessary? You and your child might enjoy trying to answer this by coloring in all the other states. Sometimes there will be a choice to make. Wyoming, for example, can be either red or green. The decision you make here will affect future choices, and if you find yourself forced to use a fifth color later, you might want to go back and do things differently. If you're careful, you should be able to complete the map with only four colors.

Mathematicians are almost certain that four colors are enough for *any* map, as long as it's drawn on a flat surface or a globe, but so far no one has ever been able to prove it. Anyone who succeeded in drawing a map that needed five colors would take his place among the great names in the history of mathematics.

The fascination of an unsolved problem is not confined to professional mathematicians; children can begin to understand it at a remarkably early age. Once they do, they will begin to appreciate mathematics for what it really is, a challenging adventure in thought.